Human Genetics

PRENTICE-HALL FOUNDATIONS OF MODERN *Genetics* SERIES

Sigmund R. Suskind and Philip E. Hartman, Editors

AGRICULTURAL GENETICS
  James L. Brewbaker

GENE ACTION, Second Edition
  Philip E. Hartman and Sigmund R. Suskind

EXTRACHROMOSOMAL INHERITANCE
  John L. Jinks

DEVELOPMENTAL GENETICS*
  Clement Markert and Heinrich Ursprung

HUMAN GENETICS, Second Edition
  Victor A. McKusick

POPULATION GENETICS AND EVOLUTION
  Lawrence E. Mettler and Thomas G. Gregg

THE MECHANICS OF INHERITANCE, Second Edition
  Franklin W. Stahl

CYTOGENETICS
  Carl P. Swanson, Timothy Merz, and William J. Young

*Published jointly in Prentice-Hall's *Foundations of Developmental Biology Series*.

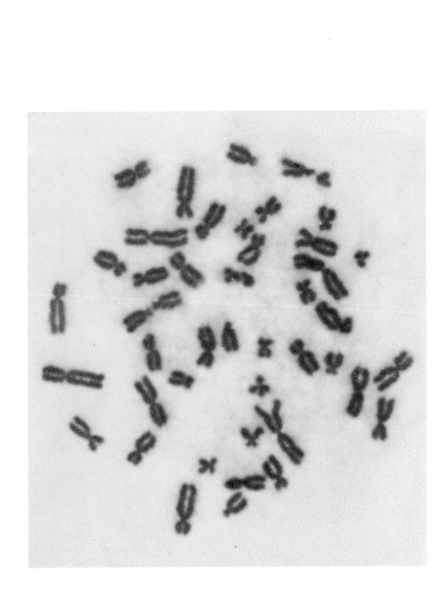

HUMAN
GENETICS

Second Edition

Victor A. McKusick
The Johns Hopkins University

PRENTICE-HALL, INC.   *Englewood Cliffs, New Jersey*

FOUNDATIONS OF MODERN GENETICS SERIES
Sigmund R. Suskind and Philip E. Hartman, Editors

Printed in the United States of America

C–13–445114–7
P–13–445106–6
Library of Congress Catalog Card Number:
69–19871

Current printing (last digit):
12   11   10   9   8   7   6   5   4   3   2

FRONTISPIECE. *Chromosomes of a single human white blood cell in the metaphase stage of mitosis.*

PRENTICE-HALL INTERNATIONAL, INC., *London*
PRENTICE-HALL OF AUSTRALIA, PTY., LTD., *Sydney*
PRENTICE-HALL OF CANADA, LTD., *Toronto*
PRENTICE-HALL OF INDIA PRIVATE LIMITED, *New Delhi*
PRENTICE-HALL OF JAPAN, INC., *Tokyo*

Foundations of Modern *Genetics*

The books in this series are intended to lead the alert reader directly into the exciting research literature of modern genetics. The forefront of genetic research draws heavily on concepts and tools of chemistry, physics, and mathematics. Because of this, the principles of genetics are presented here together with discussions of other relevant scientific areas. We hope this approach will encourage a fuller comprehension of the principles of genetics and, equally important, of the types of experiments that led to their formulation. The experimental method compels the questions: What is the *evidence* for this concept? What are its *limitations?* What are its *applications?*

Genetics today is penetrating increasingly into new areas of biology. Its rapidly expanding methodology is enabling research workers to find answers to questions that it was futile to ask only a short while ago. Even more provocative studies now underway are raising new heretofore unimagined questions.

The design of the individual short volumes of the Prentice-Hall Foundations of Modern Genetics Series permits stimulating, selective, and detailed treatments of each of the various aspects of the broad field of genetics. This facilitates more authoritative presentations of the material and simplifies the revisions necessary to keep abreast of a rapidly moving field. Each volume has its own individual focus and personality while at the same time overlapping with other volumes in the Series sufficiently to allow ready transition. Collectively, the Series, now complete, covers the main areas of contemporary genetic thought, serving as a thorough, up-to-date textbook of genetics—and, we hope, pointing the reader toward experiments even more penetrating than those described.

SIGMUND R. SUSKIND
PHILIP E. HARTMAN

vii

Preface

This survey of human genetics proceeds from the genetic material of man through the behavior of genes in kindreds, in individuals, and in populations to the implications of human genetics for human evolution, medicine, and society. I have placed strong emphasis on genetic analysis in man and I have discussed in detail the methods now available: the pedigree method, linkage analysis, the chemical structure of proteins, and cytogenetic correlations.

The frequent use of rare, often abnormal, traits to illustrate the normal mechanisms of inheritance and gene action in man is not difficult to justify. Often the genetics of rare traits is less complicated than that of normal traits. Furthermore, aberrations from normal mechanisms are frequently valuable clues to the nature of the normal physiologic mechanisms. "Treasure your exceptions!" was the sound advice given by William Bateson (1861–1926), an early student of Mendelism. In 1657 William Harvey, who discovered the circulation of the blood, expressed the same idea in espousing the study of rare diseases:

Nature is nowhere accustomed more openly to display her secret mysteries than in cases where she shows traces of her workings apart from the beaten path; nor is there any better way to advance the proper practice of medicine than to give our minds to the discovery of the usual law of Nature by careful investigation of cases of rarer forms of disease. For it has been found, in almost all things, that what they contain of useful or applicable nature is hardly perceived unless we are deprived of them, or they become deranged in some way.

There is a tendency to underestimate the usefulness of man as an object for genetic study. It is true that human genetic analysis is hampered by long generation time and small families; however, it is an advantage that anatomical, chemical, physiological, and pathological descriptions of phenotype in man are both extensive and intensive. Many physicians and workers in paramedical professions collect information of genetic significance that can be exploited by competent students of the science. Biochemical genetics was first studied in man—by Archibald Garrod (1857–1936), a London physician—in the early years of this century. Unfortunately the experience and conclusions in man did not enter the mainstream of genetic thought and theory until the 1940's. In man over sixty traits are known to be determined by genes, most of them probably nonallelic, on the X chromosome. The list represents the largest group of genes located on one chromosome in any metazoan other than *Drosophila*. *Homo sapiens* ranks with *Mus musculus, Drosophila, Zea mays, Neurospora crassa, Aspergillus nidulans, Escherichia coli*, and bacteriophage as a favorite organism for study by geneticists.

Genetics has been defined as the science of differences (or the science of variation). The outcome of matings between individuals carrying differences has been the main approach to study. Small families and the lack of controlled matings put limitations on the application of classical approaches. Molecular biology has brought about changes. Particularly the principle of the colinearity of purine-pyrimidine bases in DNA and amino acids in protein has permitted genetic inferences to be made in man from the study of proteins. Chemical methods have thus become surrogates for formal genetic analysis, which in man is otherwise awkward and crude.

In general, studies in other organisms have laid the foundation for interpretation of the genetics of man. Mainly through a combination of studies in other organisms (in which more refined genetic analysis is possible) with those in man (in which phenotypic analysis is often more detailed), human genetics has been able to make its greatest contributions to the science of genetics. The science of human heredity cannot depend, however, only on extrapolation from findings in other species. Some aspects can be studied conclusively only in man. The example of sex determination in man, which is quite different from that in many other extensively studied forms, notably *Drosophila*, leads one to agree, although for different reasons, with Pope: "The proper study of mankind is man."

Genetics is to biology what the atomic theory is to the physical sciences. Human genetics has implications not only for all aspects of the science of man, but also for the cultural, political, and social aspects of human activity. Man has profited greatly from his knowledge of the genetics of the "economic" plants, animals, bacteria, viruses, and fungi. Although control of his own genetic future to a comparable extent has not yet been feasible, an understanding of his own genetics has been notably useful in some areas. For example, the dramatic advances in surgery, such as that on the interior of the heart, would have been hampered, if not impossible, without transfusion of large volumes of blood from other individuals. Knowledge of the manifold genetic differences in blood types among individuals has made such transfusions safe and "tissue typing" is becoming a practical means of determining which specific persons can with greatest success donate organs to a specific recipient.

Great areas of ignorance about the detailed genetics of many of the important variable attributes of man, intelligence, for instance, are all too obvious. However, an optimistic attitude toward long-range accomplishments is justified on the basis of the progress made to date. I hope the reader is challenged by the gaps in our knowledge and catches from this brief survey some of the excitement of discovery in the rapidly advancing field of human genetics.

V.A.M.

Contents

Human Genetics

Historical Introduction

Long before 1900, when Mendel's observations of the 1860s were rediscovered, and even long before Mendel himself published his findings, simple patterns of inheritance in man were known, although not understood. The traits were, naturally, obvious ones such as polydactyly (extra fingers), hemophilia, and color blindness. For example, in Berlin in the 1750s Maupertuis (1689–1759) described the autosomal dominant inheritance of polydactyly and discussed segregation in terms prophetic of Mendelism. The essential features of X-linked recessive inheritance of hemophilia were described in three unrelated New England families by Otto in 1803, by Hay in 1813, and by the Buels in 1815. In 1820, Nasse, a physician in Bonn, formally outlined this pattern of inheritance, later referred to as Nasse's law. Much earlier, the Talmud made a provision for dispensing with circumcision in newborn males whose older brothers or maternal uncles had displayed a bleeding tendency. In 1876, Horner, a Swiss ophthalmologist, described the X-linked recessive pattern of color blindness.

Inheritance in a pattern we now recognize as autosomal recessive—the occurrence of a trait in multiple sibs, the offspring of normal parents, especially when the parents were related—was described in 1814 by Joseph Adams, who was far ahead of his time in his understanding of the dynamics of genetic disease. Further evidence of the importance of studying the biologic consequences of consanguineous marriages is found in Bemiss' investigation, reported to the American Medical Association in 1857.

The twin method for separating the effects of heredity and environment was suggested by Francis Galton in 1876, although at first he was not clear about the distinction between monozygotic (identical) and dizygotic (nonidentical) twins. Along with his concept of regression Galton also initiated studies of quantitative genetics, that is, polygenic inheritance.

Shortly after the rediscovery of Mendelian theory in 1900, Archibald Garrod, on the advice of William Bateson, interpreted the pattern of inheritance of alkaptonuria (a metabolic disorder in which a substance called alkapton is excreted in the urine, which turns black on standing) in Mendelian recessive terms and recognized the significance of parental consanguinity. W. C. Farabee was one of the first to trace an autosomal dominant trait (in this case, brachydactyly, or short fingers) through a family and interpret its distribution specifically in Mendelian terms. In 1911 Thomas Hunt Morgan and E. B. Wilson of Columbia University demonstrated that the characteristic pattern of inheritance of hemophilia and of color blindness is consistent with the location of the responsible genes on the X chromosome.

Man was the first organism in which biochemical genetics was studied. From studies of alkaptonuria and certain other hereditary disorders, in 1908 Garrod, the London pediatrician mentioned above, developed his inspired concept of "inborn errors of metabolism." George Beadle, who in 1958 shared the Nobel Prize in physiology and medicine for his contributions to biochemical genetics, pointed out that his one-gene–one-enzyme hypothesis was really implicit in Garrod's work and was actually formulated by Garrod in essentially the same terms.

Also in 1908, G. H. Hardy, a mathematician at Cambridge University, and Wilhelm Weinberg, a physician in Stuttgart, independently laid the foundation of population genetics with what has been designated the Hardy–Weinberg law. Again the stimulus to the development came from human genetics—a consideration of the distribution of Mendelizing traits in human populations. In the early days of Mendelism it seemed to some that a dominant trait should increase in frequency and replace the recessive trait. "Why," it was asked, "does not everyone have brachydactyly?" Hardy and Weinberg considered gene frequency (actually *gene* was not yet a generally used part of the vocabulary) the most important aspect of population genetics. They showed, furthermore, that if disturbing factors were not operating, one would expect the frequency of genes and the traits for which they are responsible to remain constant from generation to generation. Weinberg is responsible for other contributions in the statistical methodology of human genetics—the Weinberg method for estimating the proportion of twin pairs that are monozygotic, and the Weinberg method for correcting for bias of ascertainment in estimating the proportion of sibs expected to be affected by an autosomal recessive trait.

In addition to the concepts of single-factor inheritance, Mendel formulated the concept of multifactorial inheritance, that is, the additive effects of more than one gene pair. His crossing of white-flowered beans with purple-flowered beans resulted in an intermediate F_1 generation and an F_2 generation that displayed flowers ranging in color from purple to white. Early in this century a controversy raged, especially in England, between the biometricians such as Francis Galton and Karl Pearson and the exponents of Mendelism, such as Bateson, over the nature of inheritance in man. The Galton school had been studying quantitative traits such as intelligence and stature before rediscovery of Mendel's works, whereas the Mendelists in the early years after 1900 were concerned with the inheritance of discontinuous traits. The two approaches were considered incompatible until 1918, when R. A. Fisher (1890–1962) demonstrated that multiple pairs of genes, each behaving in a Mendelian manner, account for the findings of the biometricians on quantitative traits. In that discussion Fisher introduced the term *variance*.

In the 1920s and 1930s important contributions to the theory of population genetics and evolution were made by R. A. Fisher and J. B. S. Haldane in England, by Sewall Wright in this country, and by Gunnar Dahlberg in Sweden. These workers and others, such as Fritz Lenz, Lancelot Hogben, and Felix Bernstein, have contributed greatly to the statistical methodology of human genetics, for example, the methods for segregation analysis, linkage analysis, and estimation of mutation rates.

The atomic age has been accompanied by an intensified study of the genetic effects of radiation. Although precise information must come from organisms available in large numbers for investigation under controlled conditions, estimation of human mutation rates and other studies of human populations have been stimulated.

In recent years, and especially in the last decade, there has been a great increase in appreciation of the importance of genetics in understanding diseases of man. Major contributions to human genetics have come from that part of the discipline that is called *medical genetics* because of its concentration on inherited pathologic traits. In biochemical genetics the primary role of gene action in protein synthesis has been elucidated through the work of Linus Pauling, Vernon Ingram, and others, on hemoglobin variants. In 1959, by studies of the sex-chromosome constitution in sex anomalies, the mechanism of sex determination in man was shown to be different from that in *Drosophila*. Other chromosomal anomalies are being discovered; the first and most impressive of these was the one responsible for "Mongoloid idiocy" (Down's syndrome) as described by Jérôme Lejeune in 1959.

As recently as 1956 the true diploid chromosome number of man was established as 46, not 48, by J. H. Tjio and Albert Levan and by Charles Ford and John Hamerton.

The concept of genetic polymorphism, as expounded by E. B. Ford in 1940, is relatively new. Other than color blindness, however, the first discontinuous human trait that is now recognized as a genetic polymorphism was discovered by Karl Landsteiner in 1900, about the same time that Mendel's experiments were rediscovered. This was the ABO blood-group system. Since that time, 14 other major blood-group systems, one of them X-linked, have been discovered, and these systems have provided some of the clearest examples of the operation of Mendelian principles in man. Discovery in 1941 of materno-fetal incompatibility for the Rh blood groups uncovered a whole "new" category of genetically conditioned disease.

The polymorphism for ability to taste phenylthiocarbamide (PTC) was discovered in 1931. The recent search for other genetic polymorphisms of man has surged ahead through the use of diverse methods for showing immunologic, physicochemical, and enzymatic variations in proteins. The starch gel electrophoresis method that Oliver Smithies devised in 1955 has proved especially useful in revealing protein differences.

By the very definition of the term *polymorphism* (see p. 88) selection must be one of the main factors involved in the creation of polymorphisms, if not in their maintenance. The role of selection in shaping the genetic constitution of man has been studied with increasing interest since the demonstration by A. C. Allison in 1954 of the role of malaria in maintaining the high frequency of the gene for sickle hemoglobin in populations of West Africa.

The eugenics movement, which antedated the rediscovery of Mendelism, has as its laudable objective an "improvement" of the species. The movement has, however, often brought ill repute to human genetics and has, in the opinion of most, impeded the progress of genetics as a science. There is not enough scientific information on which to base recommendations for large-scale eugenic action. Perversion of eugenics in the racist philosophy of the National Socialist regime of Germany undoubtedly caused a significant setback. Despite our present knowledge of human genetics, most scientists in the field consider insight so vague—and perhaps the morality of man so underdeveloped—as to make any significant eugenic measures inordinately risky.

Seven Nobel Prizes in Physiology and Medicine have been awarded to workers in the area of genetics.

>1933 Thomas Hunt Morgan (1866–1945), for research on the nature of the gene.
>1946 Hermann Joseph Muller (1890–1967), for discovery of the induction of mutation by X ray.

1958　George Wells Beadle (1903–　) and
　　　　Edward Lawrie Tatum (1909–　), for contributions in bio-
　　　　chemical genetics; and
　　　　Joshua Lederberg (1925–　), for discovery of sexual recom-
　　　　bination in bacteria.

1959　Arthur Kornberg (1918–　) and
　　　　Severo Ochoa (1905–　), for studies of the chemistry of
　　　　DNA and RNA.

1961　James D. Watson (1928–　),
　　　　Francis H. C. Crick (1916–　), and
　　　　Maurice H. F. Wilkins (1916–　), for elucidation of the
　　　　intimate structure of DNA.

1965　François Jacob (1920–　),
　　　　André Lwoff (1912–　), and
　　　　Jacques Monod (1910–　), for their discoveries concerning
　　　　genetic control of enzyme and virus synthesis.

1968　Marshall W. Nirenberg (1927–　),
　　　　H. G. Khorana (1922–　), and
　　　　Robert W. Holley (1922–　), for "cracking the genetic
　　　　code" and elucidating the means by which a gene determines
　　　　the sequence of amino acids in a protein.

By way of summary, my personal list of the 12 most significant discoveries in human genetics follows. These are discoveries that were
made predominantly in man but are applicable in other animals.

(1) The twin method was first developed by Galton in 1876.

(2) In the early years of this century Garrod focused attention on
genetic blocks in chains of metabolic reactions.

(3) In 1908, Hardy and Weinberg focused attention on gene frequency as the pertinent variable in population genetics. They elab-
orated the fundamental principle on the basis of which factors leading
to change in gene frequencies could be analyzed.

(4) Landsteiner (Nobel laureate in 1930) initiated the study of
simply inherited polymorphic traits in man by his discovery in 1900 of
the ABO blood-group system. (Color blindness was, of course, discovered earlier.)

(5) The role of infectious disease in shaping the genetic constitution of man was suggested by E. B. Ford, J. B. S. Haldane, and others.
The first clear evidence for such a role was provided by A. C. Allison in
1954, in his study of the relationship of malaria and the gene for sickle
hemoglobin.

(6) J. B. S. Haldane's estimate in 1937 (revised in 1947) of the
distance separating the color blindness and hemophilia loci on the X
chromosome was the first linkage studied quantitatively in man. Jan
Mohr's demonstration (1951, 1953) of the linkage between Lutheran
blood group and secretor factor was the first discovery of an autosomal
linkage in man.

(7) In 1952, Gerty T. Cori and Carl F. Cori demonstrated the first
specific enzyme defect in an inborn error of metabolism: deficiency of

glucose-6-phosphatase in a form of glycogen-storage disease. In 1953, George Jervis identified deficiency of phenylalanine hydroxylase in phenylketonuria (PKU). Garrod's prediction of an enzyme defect in alkaptonuria was confirmed by Bert La Du and his colleagues in 1958.

(8) The role of the gene in determining the amino acid sequence of a protein was discovered in 1957 by Vernon Ingram, who studied the difference between normal hemoglobin A and sickle hemoglobin of man.

(9) In 1959, Lejeune and his colleagues first discovered a chromosomal aberration to be the basis of a congenital malformation in man —Mongoloid idiocy (also called "Down's syndrome") in this case.

(10) The role of the Y chromosome in sex determination in man was discovered by C. E. Ford and P. A. Jacobs and their colleagues in 1959.

(11) In 1959, the first specific chromosomal aberration associated with malignant neoplastic disease of man was discovered in a form of leukemia by Peter Nowell and David Hungerford (see p. 31).

(12) In 1961 several workers simultaneously suggested that one X chromosome in the normal female is relatively inactive genetically (the Lyon hypothesis). See page 103 for further explanation.

Exciting new developments in our understanding of the genetics of man continue, with no sign of waning. Studies of the meiotic chromosomes in both the male and the female promise to provide insight beyond that afforded by the mitotic chromosomes. Phenomenal genetic diversity is revealed by the biochemical and serologic polymorphisms which are being discovered in large numbers. The rapidly enlarging list of polymorphisms demands of the population geneticist an explanation of the forces which establish and maintain them, and provides the student of linkage with markers for mapping the chromosomes of man. Largely deprived of the conventional genetic methods of the experimental geneticist, the human geneticist has been clever in devising substitutes. The principle of colinearity of nucleotides in DNA and amino acids in the product protein has permitted inferences about the fine structure of genes to be made from protein analysis. Cell culture methods not only provide an excellent approach to elucidation of the basic defect in inborn errors of metabolism but, through cell hybridization, genetic conclusions such as the localization of specific genes to specific chromosomes can also be made. Immediate applications of advances in human genetics are evident in the use of tissue typing for matching donor and recipient in organ transplantation, amniocentesis with study of amniotic cells for antenatal diagnosis of inborn errors of metabolism or chromosomal aberrations and in the treatment of some hereditary disorders. The above enumeration will undoubtedly look pale by comparison with the progress of the next few decades. Past experience would suggest that a combination of ingenious method and incisive thinking will be the key to advancement.

References

Boyer, S. H., IV, ed., *Papers on Human Genetics.* Englewood Cliffs, N.J.: Prentice-Hall, Inc., 1963.

Carlson, E. A., *The Gene: A Critical History.* Philadelphia: W. B. Saunders Co., 1966.

Dunn, L. S., *A Short History of Genetics: Development of Some of the Main Lines of Thought.* New York: McGraw-Hill Book Company, 1965.

Garrod, Archibald E., *Inborn Errors of Metabolism.* Reprinted with a supplement by Harry Harris. London: Oxford University Press, 1963.

McKusick, Victor A., "Walter S. Sutton and the Physical Basis of Inheritance," *Bull. Hist. Med., 34* (1960) , 487–97.

——, "Hemophilia in Early New England," *J. Hist. Med., 3* (1962) , 342–65. A follow-up of four kindreds in which hemophilia occurred in the pre-Revolutionary period.

Motulsky, A. G., "Joseph Adams (1756–1818) , a Forgotten Founder of Medical Genetics," *AMA Arch. Intern. Med., 104* (1959) , 490.

Pearson, Karl, *The Life, Letters and Labours of Francis Galton,* 4 vols. London: Cambridge University Press, 1914–30.

Penrose, L. S., "The Influence of the English Tradition in Human Genetics," in *Proceedings of the Third International Congress of Human Genetics (Sept. 1966)* , J. F. Crow and J. V. Neel, eds. Baltimore: Johns Hopkins Press, 1967.

Stern, C., and Sherwood, E. R., eds., *The Origins of Genetics; a Mendel Source Book.* San Francisco: W. H. Freeman & Co., 1966.

Sturtevant, A. H., *A History of Genetics.* New York: Harper & Row, Publishers, 1965.

Watson, J. D., *The Double Helix.* New York: Atheneum Press, 1968.

The Chromosomes of Man

In man as in other organisms, the hereditary material, DNA, is carried by the chromosomes. Somatic (body) cells, which have two of each type of chromosome, are diploid. In mitosis each chromosome is replicated and is represented in each of the two daughter cells. Gametes, or germ cells, which have one of each type of chromosome, are haploid and are produced in the process of meiosis. Segregation, assortment, and recombination occurring in the meiotic process are the basis of the characteristics of inheritance.

Until 1956 the diploid number of chromosomes in man was thought to be 48 instead of 46, and even the sex-chromosome constitution was in doubt. Improved methods for studying the mitotic chromosomes in somatic cells of man, however, have permitted a more accurate and more detailed description of the normal karyotype and the discovery of abnormalities responsible for certain congenital malformations. Two techniques in particular have helped greatly: (1) treatment of cell cultures with colchicine resulting in accumulation of dividing cells in metaphase, thereby assuring an adequate number of cells at a stage of division satisfactory for studying chromosomes; (2) treatment with a hypotonic solution producing swelling of the cells and spreading of the chromosomes, thus facilitating their study. (See *Cytogenetics,* by C. P. Swanson, T. Merz, and W. J. Young, in this series, for general background on many aspects of this chapter.)

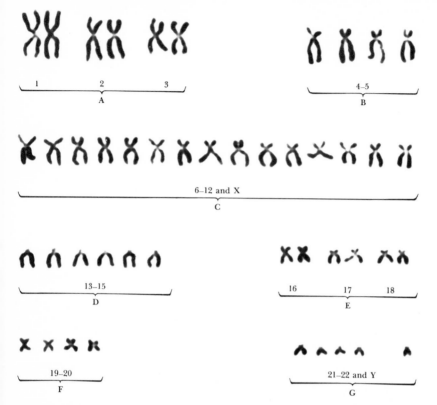

FIG. 2.1. *The mitotic metaphase chromosomes of a somatic cell of a male, arranged in a karyotype. The frontispiece is a photomicrograph of such chromosomes.*

The cells used in these studies are obtained from bone marrow (aspirated from the sternum, for example), from bits of skin or fascia removed by biopsy, or from the leukocytes of a sample of venous blood drawn by the ordinary method of venipuncture. This last method, because of its simplicity, has obvious advantages. A step that has contributed to the success of the useful technique of blood culture is the addition of phytohemagglutinin. Derived from a bean, this material produces agglutination of erythrocytes, thereby facilitating the separation of the leukocytes. In addition, phytohemagglutinin has associated with it a substance that stimulates cell division in lymphocytes.

Stains for the DNA of chromosomes, such as aceto-orcein or Feulgen's are applied, and the preparations are examined by ordinary light microscopy. Individual chromosomes are cut out from enlarged photographs, matched in pairs of homologous chromosomes, and arranged in order of descending length (Figs. 2.1 and 2.2).

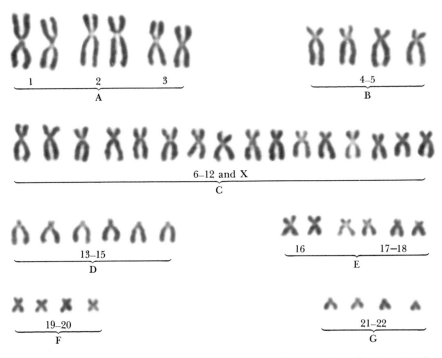

FIG. 2.2. *The mitotic metaphase chromosomes of a somatic cell of a female, arranged in a karyotype.*

The Normal Chromosome Constitution

The normal diploid chromosome number of man is 46. The sex-chromosome constitution of the male is XY, and of the female XX. In addition there are 22 pairs of autosomes. In metaphase of mitosis each chromosome consists of two identical chromatids that separate in later phases, each to become one of the 46 chromosomes of one of the two daughter cells. The two chromatids are joined at the *centromere* (also called *kinetochore*, or *primary constriction*).

In man three classes of chromosomes (Fig. 2.3) are recognized according to position of the centromere and the resulting relative length of the arms, that is, the parts of the chromosome on each side of the centromere. The three classes are: (1) median, or metacentric—centromere in an approximately central position with arms of equal length; (2) submedian, or submetacentric—centromere nearer one end than the other, resulting in one short arm and one long arm; and (3) acrocentric ("extremity center"), or subterminal—centromere near the end so that one arm is very short.

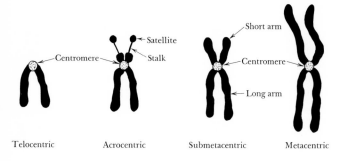

Telocentric   Acrocentric   Submetacentric   Metacentric

FIG. 2.3. *Types of metaphase chromosomes. Note the satellites on the acrocentric chromosome. No chromosome of man is telocentric.*

Placing the chromosomes in a karyotype is at best an approximation. Especially in Group C, comprising 6–12 plus X, and others such as Group G (21 and 22), one cannot be absolutely certain that the chromosomes that seem identical on the basis of length and arm ratio are in fact homologous. Several chromosomes can equally well be paired; moreover, there is no certainty that what is, for example, called chromosome 10 in one cell is the same as what is designated chromosome 10 in another. The chromosomes can, however, be put into groups as shown in Table 2.1.

The two sex chromosomes of the female are identical and are referred to as the X *chromosomes*. The X chromosome has a submedian centromere, corresponds to Group C of the autosomes, and is usually pair number 7 in terms of total length. The male has one X chromo-

TABLE 2.1. *Human Chromosome Analysis by Groups*

Group	Size and centromere position	Ideogram number	Number in diploid cell
A	Large; median/submedian	1–3	6
B	Large; submedian	4, 5	4
C	Medium; submedian	6–12 and X	15 (male) or 16 (female)
D	Medium; subterminal	13–15	6
E	Small; median/submedian	16–18	6
F	Smallest; median	19, 20	4
G	Small; subterminal	21, 22, and Y	5 (male) or 4 (female)

some and a small acrocentric chromosome, the Y chromosome, that sometimes cannot be differentiated with certainty from the other four acrocentric chromosomes of the male. The Y chromosome tends to vary, especially in length, from male to male, with close similarity in males of any one family.

Another distinguishing feature of the normal chromosomes is the presence of satellites, which are chromatin knobs connected to the short arm of certain chromosomes by a stalk, or secondary constriction. Satellites are found attached to five pairs of acrocentric chromosomes: chromosomes 13, 14, and 15 (Group D) and chromosomes 21 and 22 (Group G). However, ten satellited chromosomes are almost never identified in a single cell.

Secondary constrictions have also been identified in the long arms of chromosomes 1, 9, and 16, and occasionally others. These constrictions also assist in the identification of individual chromosomes. Sites of secondary constriction participate in the organization of the nucleolus, which serves some function in connection with synthesis of ribosomal RNA, a constituent of the cytoplasmic organelle, the ribosome, on which protein is synthesized.

Some chromosomes can be distinguished through autoradiography by a characteristic pattern in which new DNA is synthesized. Dividing cells incorporate the radioactive nucleoside, tritium-labeled thymidine, into the DNA of chromosomes.

Normal variation of the chromosomes occurs in some instances and often can be shown to be transmitted from generation to generation. Chromosome 1 is sometimes unusually long because of an "uncoiled" segment (see p. 73 and Fig. 2.4). Chromosomes 16 and Y are notably

FIG. 2.4. *The pair of No. 1 chromosomes in* (a) *a person showing an "uncoiled" segment in the long arm of one chromosome 1 and* (b) *his niece who did not inherit the anomalous chromosome.*

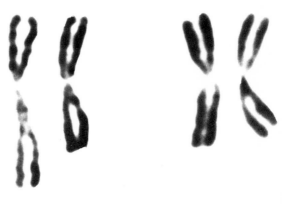

(a)   (b)

variable in length, although the ease of recognizing variation in these may account for an exaggerated idea of the frequency relative to that of variation in other chromosomes. The deletion of the short arm of one Group G chromosome has been observed as an apparently normal familial characteristic. A "giant satellite" also occurs as a familial trait (see p. 76 ff).

The detailed structure of the chromosomes will not be discussed in this book. The part of the chromosome which carries the "genetic information" is, of course, now well established to be deoxyribonucleic acid (DNA).

Sex Determination in Man

In the late 1940s Murray Barr discovered a difference in the interphase somatic nuclei of males and females: a chromatin mass called the sex chromatin, or Barr body, is present in the normal female, but not in the normal male. This discovery and others in cases of sex anomaly were the beginning of our present understanding of the mechanism of sex determination in man and other mammals.

The Barr body (Fig. 2.5) can be identified in a significant proportion of cells in all tissues of the human female, and there is reason to think that it is present in essentially all female somatic cells, at least at some stage of interphase. Furthermore, in the human female the polymorphonuclear leukocytes tend to show "drumsticks," a characteristic pedunculated lobule of the nucleus (see Fig. 2.6). The simplest method for determining sex chromatin status (and therefore the method used in surveys) makes use of epithelial cells obtained by scraping the lining of the cheek with a tongue depressor. Thus, the XX sex-chromosome constitution of the normal female and the XY constitution of the normal male can be determined directly from karyotype analysis of somatic cells in metaphase (for example, see Figs. 2.1

FIG. 2.5. *The Barr body, or sex chromatin.* (a) *Cells of the normal female have one Barr body as indicated by the arrow.* (b) *Cells of the normal male lack the Barr body.* (c) *Persons with three X chromosomes (the XXX or XXXY syndrome) have two Barr bodies.*

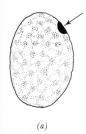  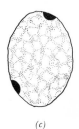

(a)   (b)   (c)

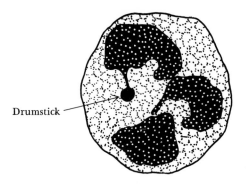

Drumstick

FIG. 2.6. *The "drumstick" that can be identified in about 5 percent of the polymorphonuclear leukocytes of the normal female.*

and 2.2), or indirectly from the sex chromatin and the leukocyte drumsticks.

But what is the critical difference between the normal male and female? Is the human male a male by default? Does the presence of two X chromosomes determine a female and the presence of one X chromosome determine a male? Or does the Y chromosome play an active role in determining maleness? The answers to these questions have come largely from the study of sex anomalies.

When the Barr technique was applied to sex anomalies of man it was discovered that certain persons had sex chromatin findings inappropriate to the phenotype. Specifically, persons with the Turner syndrome have external genitalia of female type but are chromatin negative, and persons with the Klinefelter syndrome are chromatin positive despite external genitalia of male type. (Examples of these two anomalies are shown in Figs. 2.7 and 2.8, with a description of the phenotypic features.) The sex chromatin findings suggest a sex-chromosome abnormality, which is indeed found on study of the metaphase chromosomes: in most cases of the chromatin-negative Turner syndrome there are only 45 chromosomes and only one sex chromosome, an X. The sex-chromosome constitution is said to be XO ("X-Oh"). In the chromatin-positive Klinefelter syndrome 47 chromosomes are found, there being two X chromosomes and a Y. The sex-chromosome constitution is said to be XXY.

A chromosomal aberration characterized by a deviation from the normal total number that is not a multiple of the normal chromosome number (say, 45 chromosomes or 47 chromosomes) is known as aneuploidy. Polyploidy exists when the chromosome number is some simple multiple of the normal haploid number. For example, many liver cells are tetraploid, with 92 chromosomes.

The Klinefelter syndrome occurs once in every 400 to 600 "male" births; the Turner syndrome is less frequent, occurring about once in each 3,500 "female" births. (A majority of XO conceptions end in spontaneous abortions; see Table 2.3.) There are several theoretically

FIG. 2.7. *The Turner syndrome.* (a) *The features are female external genitalia, short stature, webbed neck, low-set ears and typical facies, broad shield-like chest with widely spaced nipples and undeveloped breasts, small uterus, and ovaries represented only by fibrous streaks. In some cases coarctation of the aorta (a marked narrowing just beyond the mouth of left subclavian artery) leads to severe hypertension in the upper part of the body. Such has been corrected surgically in the patient illustrated here; note the surgical scar on the left side of the thorax. The patient is "chromatin-negative."* (The scale of this and similar pictures which follow is in centimeters.) (b) *The karyotype of this patient, with 45 chromosomes and an XO sex-chromosome constitution. This photo appeared in* J. Chronic Diseases (*July 1960*) *and in* Medical Genetics 1958–1960 (*St. Louis, Mo.: C. V. Mosby Company, 1961*); *reproduced by permission of the editors of the* Journal of Chronic Diseases.

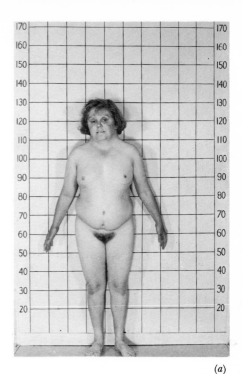

(a)

1   2   3
A

4–5
B

6–12 and X
C

13–15
D

16   17–18
E

19–20
F

21–22
G

(b)

15

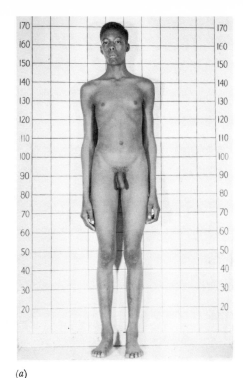

(a)

FIG. 2.8 *The Klinefelter syndrome.* (a) *The external genitalia are of male type but the testes are consistently very small and body hair is sparse. Most of the cases have gynecomastia—female-like breast development. Patients tend to be unusually long-legged. They are "chromatin positive."* (b) *The karyotype of this patient, with 47 chromosomes and an XXY sex-chromosome constitution. This photo appeared in* J. Chronic Diseases *(July 1960) and in* Medical Genetics 1958–1960 *(St. Lous, Mo.: C. V. Mosby Company, 1961); reproduced by permission of the editors of the* Journal of Chronic Diseases.

(b)

| 1 | 2 | 3 |
| A | | |

| 4–5 |
| B |

| 6–12 and X |
| C |

| 13–15 | | 16 | 17 | 18 |
| D | | E | | |

| 19–20 | | 21–22 and Y |
| F | | G |

16

possible mechanisms by which these conditions might occur, and there is precedence in other organisms for suspecting the accident of non-disjunction either in one of the two meiotic divisions of gametogenesis in one of the parents or in the early mitotic divisions of the zygote. Figure 2.9 presents these mechanisms schematically.

Study of X-chromosome marker traits in the afflicted person and in both parents provides some indication of the origin of the X chromosomes in XO and XXY cases (Fig. 2.10). Color blindness has, for example, been observed in XO individuals whose parents both have

FIG. 2.9. *Mechanisms by which aneuploidy of the sex chromosomes might develop.* (a) *Nondisjunction in gametogenesis, which can occur either at the first meiotic division, as shown here, or at the second divison.* (b) *Nondisjunction or chromosome loss in the zygote. Note the mechanism by which the nonmosaic XXY Klinefelter syndrome might arise from an XY zygote.*

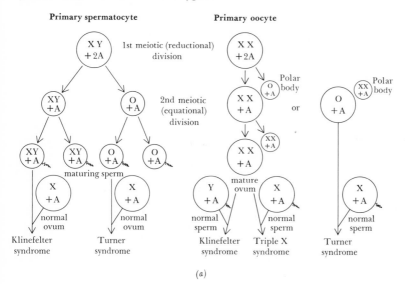

(a)

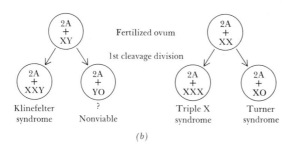

(b)

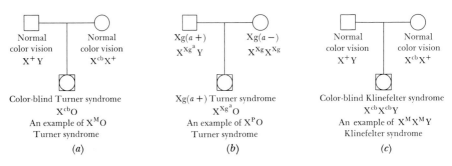

FIG. 2.10. *Demonstration of the X^{M}O and X^{P}O states and of the X^{M}X^{M}Y state by family studies.*

normal color vision (Fig. 2.10a). Since color blindness is an X-linked recessive trait, it is concluded that the mother is a heterozygous carrier for color blindness, that she contributed to the offspring the X chromosome bearing the mutant allele for color blindness, that the case of Turner syndrome is of the X^{M}O type, and that it is the paternal sex chromosome that is missing. Either the ovum was fertilized by a sperm without any sex chromosome or the paternal sex chromosome, either X or Y, was lost at an early stage, perhaps in the rearrangement of the male pronucleus that occurs in the time between the entrance of the sperm into the egg and completion of fertilization. Cases of X^{P}O Turner syndrome have been identified by means of the blood type Xg^a, which is an X-linked dominant (Fig. 2.10b). In these cases the maternal X chromosome is missing. The origin of the XXY Klinefelter syndrome has been investigated by a similar approach (Fig. 2.10c).

Other sex-chromosome anomalies have been discovered in recent years. One of these, phenotypically female, has two Barr bodies in the nuclei of buccal scrapings, and a sex-chromosome constitution of XXX. Another of these, phenotypically male, has two Barr bodies and a sex-chromosome constitution of XXXY. Other anomalies as indicated in Table 2.2 have been recognized. Many of these patients are mentally retarded and have been detected by sex-chromatin surveys in institutions for the retarded. The brain is such a delicately balanced mechanism that it is most vulnerable to be thrown out of kilter by chromosomal aberrations.

At least two principles emerge from the data outlined in Table 2.2.

(1) The maximum number of Barr bodies in any one cell is one less than the number of X chromosomes.

(2) The male sex phenotype is precisely correlated with the presence of a Y chromosome. In the absence of a Y chromosome the sex phenotype develops along female lines regardless of the number of X chromosomes present.

The mechanism of sex determination in man and in the mouse is

TABLE 2.2. *Sex Determination*

	Sex phenotype	Fertility	Number of Barr bodies	Sex-chromosome constitution
Normal male	male	+	0	XY
Normal female	female	+	1	XX
Turner syndrome	female	−	0	XO
Klinefelter syndrome	male	−	1	XXY
XYY syndrome	male	+	0	XYY
Triple X syndrome	female	±	2	XXX
Triple X-Y syndrome	male	−	2	XXXY
Tetra X syndrome	female	?	3	XXXX
Tetra X-Y syndrome	male	−	3	XXXXY
Penta X syndrome	female	?	4	XXXXX

quite different from that in *Drosophila* and many other nonmammalian forms in which the role of the Y chromosome is not the active one it is in most mammals. The mechanism in man parallels that in the plant *Melandrium,* where the Y chromosome has an active role in determining maleness as it does in man.

Origin of the Barr Body

Further evidence that the Barr body is derived from *one* X chromosome is provided by two observations. (1) Cells in an early phase of mitotic division show that one X chromosome of the female stains differently from the other X and from the autosomes (Fig. 2.11a); this X chromosome is said to be heterochromatic. (2) In cultures of cells from the female, thymidine labeled with tritium (H^3) can be added. The thymidine becomes incorporated into newly synthesized DNA of such cells. Asynchrony in DNA synthesis by the two X chromosomes can be observed by means of autoradiographs made of cells dividing in these cultures (Fig. 2.11b). The heterochromatic X chromosome synthesizes DNA late in the process of mitosis and is the X chromosome that constitutes the Barr body. This statement is based on the fact that the cells of persons with three X chromosomes (XXX or XXXY) have two heterochromatic X chromosomes, two late-labeling X chromosomes, and two Barr bodies; persons with four X chromosomes (XXXX and XXXXY) have three; and so on.

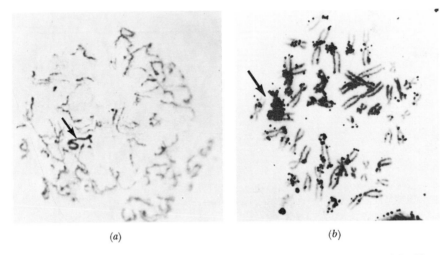

(a)   (b)

FIG. 2.11. (a) *Heterochromatic X chromosome in prophase. Courtesy of S. Ohno.* (b) *Late-labeling X chromosome in autoradiographs using tritium-labeled thymidine. Courtesy of J. L. German, II.*

Information from other species suggests that heterochromatin is genetically inactive. The Lyon hypothesis, which will be discussed in Chap. 4, proposes that the Barr body is related to dosage compensation in man and other mammals, that is, that it provides an explanation why the normal female with a double dose of X chromosome genes shows no greater effects than the normal male with a single dose. Other genetic observations are explained by the Lyon hypothesis. It should be noted that the X chromosomes in the germ-cell line of the female, that is, the two X chromosomes of the oogonia, do not display this cytologic differentiation (p. 37) .

The XYY Male

An important finding is that certain unusually tall and unusually aggressive males have 47 chromosomes and an XYY sex chromosome constitution. Such individuals often have been found in penal institutions. Many but not all are mentally retarded to some extent. Since the normal XY male is on the average taller and more aggressive than the normal XX female, the findings in XYY males may represent a dosage phenomenon resulting from the extra Y chromosome. The testes are apparently normal in these cases, and XYY males have fathered children. No instance of an XYY male with an XYY son is known. W. Court Brown suggests that finding the correlation between the XYY karyotype and disturbances of behavior "may be the most important discovery yet made in human cytogenetics, [because] it may provide a powerful lever to open up the study of human behavioural genetics." The exact frequency of the XYY male is not yet known. It is, however, probably no rarer than 1 in 2,000 male births. The rest of the genetic constitution and the environment undoubtedly modify the phenotype of the XYY male. Some XYY males are of normal intelligence and achieve acceptable social adjustment.

Abnormalities of the Autosomal Chromosomes

The first autosomal abnormality to be described in man was the one responsible for Mongoloid idiocy, or Down's syndrome. (See Fig. 2.12a for a description of this disorder.) All patients with this characteristic phenotype have all or most of chromosome 21 triply represented rather than doubly (Fig. 2.12b); in most cases there is simple trisomy of chromosome 21, each cell containing three chromosomes 21 rather than two. It is impossible to be certain whether the trisomy involves chromosome 21 or 22 since they are morphologically identical, but it is reasonable to assume that only one of these pairs is involved and that in all cases it is the same pair. By convention and for convenience the twenty-first pair is considered the affected one.

Nondisjunction comparable to that described in the sex chromosomes appears to be the basis for the anomaly in Down's syndrome. Older mothers are much more likely to have children with this condition than younger mothers (Fig. 2.13). There is indirect evidence that the nondisjunction may occur either in the ovum during meiosis or in the early cleavage stages of the zygote; possibly the former is more frequently the case. If the latter possibility occurs, then the corresponding monosomic cells (cells with only one chromosome 21) probably die. (The Turner syndrome is a monosomic state involving the X chromosome; however, no monosomic state involving autosomes has yet been discovered. It is apparently more disruptive to have too few autosomes than it is to have too many. In the X chromosome, special mechanisms of dosage compensation have developed, and monosomy or polysomy are relatively well sustained.

Translocation involving chromosome 21 and either chromosome 14 or chromosome 15 is another mechanism for Down's syndrome. Such patients are found to have 46 chromosomes with two normal chromosomes 21, one normal chromosome 14 (or 15) and an unpaired large chromosome which is interpreted as a fusion of the long arms of chromosomes 21 and 14 (or 15). In such cases of "translocation Down's syndrome," as in the more usual instances of trisomy 21, the genetic material of chromosome 21 is present in essentially triple dosage (Fig. 2.14a). Persons with translocation Down's syndrome are phenotypically indistinguishable from those with the more usual trisomy 21.

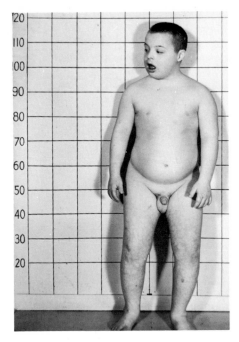

(a)

FIG. 2.12. *Down's syndrome*. (a) *The clinical features are mental retardation, a peculiarity in the folds of the eyelids suggesting the eyes of Mongoloid peoples (although in fact quite different), short stature, stubby hands and feet, peculiarity of the palm prints, and congenital malformations, especially of the heart.* (b) *The karyotype of a patient with Down's syndrome. It is uncertain whether the extra chromosome is a chromosome 21 or 22; by convention it is designated 21. This photo appeared in* Medical Genetics 1958–1960 (*St. Louis, Mo.: C. V. Mosby Company, 1961*); *reproduced by permission of the publisher.*

1	2	3		4–5
	A			B

6–12 and X
C

13–15		16	17–18
D			E

19–20		21–22 and Y
F		G

(b)

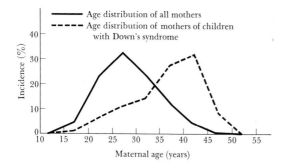

FIG. 2.13. *Age distribution of mothers of Down's syndrome patients compared with that of all mothers. Based on data of L. S. Penrose.*

When the parents of patients with translocation Down's syndrome are studied, it is sometimes found that one parent, usually the mother, is a carrier of a "balanced translocation." She is phenotypically normal but has only 45 separate chromosomes (Fig. 2.14*b*)—one chromosome 21, one chromosome 14 (or 15), and a large translocation chromosome composed of most of a chromosome 21 and a chromosome 14 (or 15). The parent is phenotypically normal because the genetic material is present almost in full and certainly not in excess amounts.

It has been found that Down's syndrome occurs in less than one-fifth of the children of females who carry the 15–21 translocation or fusion chromosome. The explanation for this proportion of affected offspring may be that schematized in Fig. 2.15. The translocation chromosome may arise by centric fusion so that the centromere of both chromosomes is represented in part or in full. Synapsis in the first stage of meiosis and the types of segregation may be diagrammed as shown in Fig. 2.15. If the three types of segregation occur with equal frequency, six types of zygotes would be expected. Of these, three are lethal. That the Down's syndrome is not present in one-third of offspring may be due in part to the loss of some through spontaneous abortion (see p. 30). If alternate segregation, the third type in the diagram, is favored, an explanation for the low proportion of Down's syndrome cases would be provided.

In connection with the above explanation, the important observations are that (1) autoradiography shows the D chromosome involved in the translocation to be either 14 or 15, not 13; (2) trisomy 14 and trisomy 15 have not been observed in nonmosaic, viable individuals; and (3) viable monosomy 21 seems to be very rare. The fact that monosomy 21 has not been observed in translocation Down's syndrome families supports the view that this is usually a lethal karyotype.

Cases of Down's syndrome with translocation of chromosome 21 to other chromosomes, for example, to chromosome 22, have also been

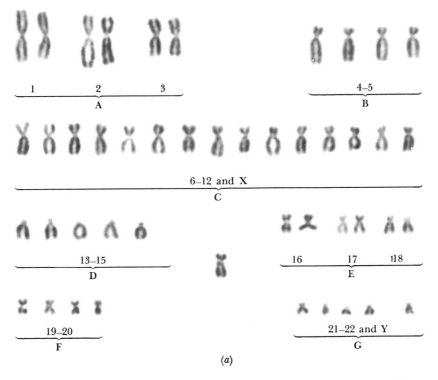

(a)

FIG. 2.14. (a) *The karyotype of a patient with translocation Down's syndrome. Forty-six chromosomes are present with one normal chromosome 15 and two normal chromosomes 21, but in addition there is a large chromosome formed by transloca-*

observed. Furthermore, some cases appear to have an isochromosome 21, that is, a chromosome consisting of the long arm of chromosome 21 in duplicate. Note that a person who although phenotypically normal has 45 chromosomes, one being a translocation or fusion chromosome involving two chromosomes 21, has a 100 per cent risk that liveborn children will have Down's syndrome. Monosomy 21 usually leads to abortion.

Female Down's syndrome patients have had offspring. The ova produced by a woman with the ordinary trisomy 21 Down's syndrome can be expected to be of two types with equal frequency: those that have one chromosome 21 and those that have two chromosomes 21. After fertilization an ovum of the latter type will develop into an individual with Down's syndrome. In keeping with these expectations, both normal and Down's syndrome offspring have been observed in about equal frequencies from mothers with Down's syndrome. If one monozygotic twin has Down's syndrome, then the other is expected

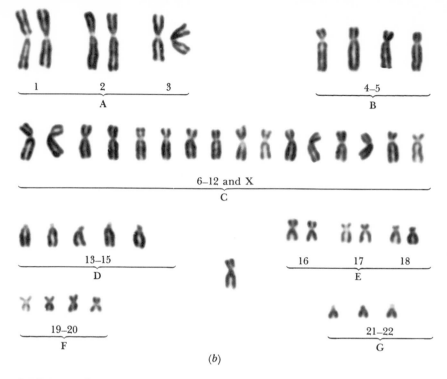

(b)

tion between chromosomes 15 and 21. (b) *Karyotype of this patient's mother, who is phenotypically normal but has 45 chromosomes, one of which is a fused chromosome 15–21.*

to have it. Experience bears out expectation; probably almost all monozygotic twins are concordant for this trait, whereas very few dizygotic twins are concordant.

Down's syndrome is by no means rare. It occurs once in each 500 or 600 births, being more frequent when mothers are older than the average. Down's syndrome is the most frequent single definable entity causing severe mental deficiency.

It is undoubtedly significant that no trisomy of the largest autosomes has been described and that the chromosome involved in Down's syndrome is one of the smallest. The trisomies of two slightly larger chromosomes, number 13 and number 18, lead to early death. The large chromosomes contain so much genetic information that the effects of overdosage are probably always lethal. Furthermore, it is noteworthy that autosomal monosomy, even for the smallest chromosomes, is very rare.

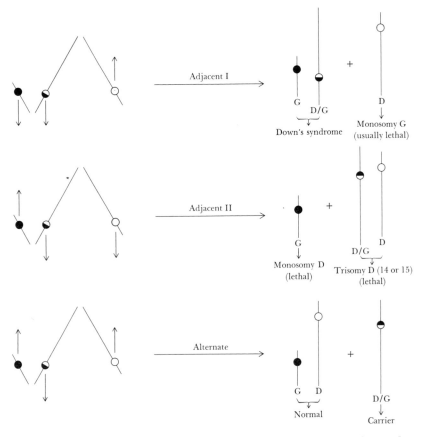

FIG. 2.15. *Schema to predict the gametic products produced by a carrier of the 15–21 fusion (translocation) chromosome.*

Mosaicism

A mixture of cells with different karyotypes is known as *mosaicism*. Some cases of the Turner syndrome have mosaicism of XO and XX cells, and some rare cases of Down's syndrome have trisomy 21 in some cells and a normal karyotype in others. Sometimes the karyotypes in mosaics are complementary, for example, XO in some cells and XXX in other cells. Such mosaicism indicates that the accident of cell division took place in the zygote (or later) rather than in gametogenesis. In a few instances, genetic markers have shown that mosaicism probably arose by double fertilization, that is, by fertilization of two eggs or an egg and a polar body by two spermatozoa. In one case, for example, the individual was XX in some cells, XY in others, and had in the case of some markers inherited both alleles from both the father and the mother.

Chimerism, a phenomenon similar to, but distinct from, mosaicism, occurs in dizygotic twins through exchange of blood cells *in utero* (see Chap. 5). The distinction between mosaicism and chimerism is that the former arose from a single genotype. The distinction is admittedly blurred in cases of double fertilization, referred to as mosaics.

Abnormalities of Chromosome Structure

An isochromosome of the long arm of the X chromosome has been observed in some cases of chromatin-positive Turner syndrome. These persons have clinical features like those of the more frequent chromatin-negative XO Turner syndrome, but have only one normal X chromosome. The other X chromosome is replaced by a larger

FIG. 2.16a. *Karyotype of patient with the Turner syndrome, showing an isochromosome of the long arm of an X chromosome* (arrow).

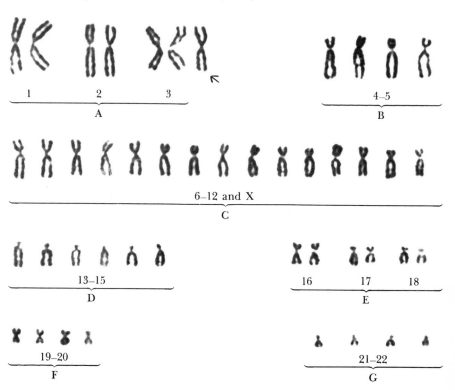

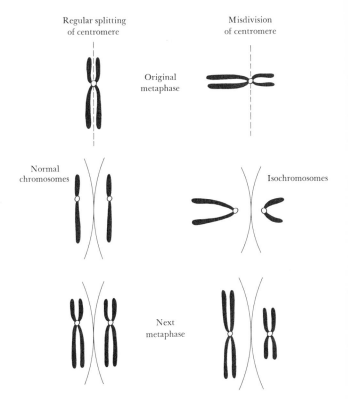

Regular splitting of centromere

Misdivision of centromere

Original metaphase

Normal chromosomes

Isochromosomes

Next metaphase

FIG. 2.16b. *The mechanism by which an isochromosome is thought to be produced.* After **D. G. Harnden, in Chromosomes in Medicine, J. L. Hamerton, ed. London: Wm. Heinemann, 1962.**

chromosome (the size of a number 3 chromosome) with a median centromere (Fig. 2.16a); this larger chromosome is thought to arise through transverse rather than longitudinal splitting of the centromere at the second meiotic division (Fig. 2.16b). The long-arm chromatids remain connected at the centromere and form the isochromosome. That the long chromosome indeed consists of duplicated long arms is supported by the identical pattern of autoradiographic labeling when tritium-labeled thymidine is added to cultures of cells in such cases. The patient has unusually large Barr bodies, and is trisomic for the long arm but deficient for the short arm of the X chromosome. Because of this deficiency many of the features of the usual XO Turner syndrome are present. A presumed isochromosome of the long arm of chromosome 21 has been found in some cases of Down's syndrome.

Deletion of part of one X chromosome has been observed as the basis of some sex anomalies. Deletion of the short arm results in a

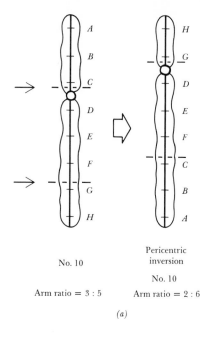

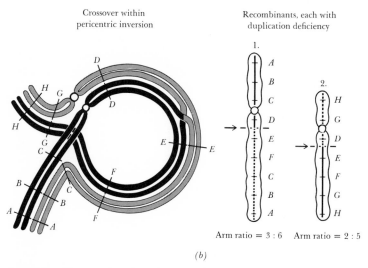

FIG. 2.17. *Diagram illustrating the origin of recombinant chromosomes as a result of crossing over within a pericentric inversion of chromosome 10. (a) The origin of the inversion. (b) Crossing over within the loop formed by the inverted segment at meiotic pairing, with the resulting recombinants.* **Courtesy of Malcolm A. Ferguson-Smith, Proc. Third Intern. Congr. Human Genetics (1966). Baltimore, Md.: Johns Hopkins Press, 1967.**

clinical phenotype essentially identical to the ordinary XO Turner syndrome. As in the case of the isochromosome Turner syndrome (**XX**), the patient is chromatin-positive; however, in the deleted-X cases (Xx), the Barr body is unusually small.

Deletion of part of the short arm of chromosome 5 leads to the *cri du chat* syndrome, which was discovered by Lejeune and so named by him because, in addition to small head, wide spacing of the eyes and mental retardation, the plaintive cry of the affected infant resembles that of a cat. (Translocation carriers who give rise to offspring with the *cri du chat* syndrome have been described. In the same family, children abnormal by reason of duplication of part of chromosome 5 have been found.) Ring chromosomes, which originate by deletion of both ends and union of the two ends of the centric piece, have been found in cases of malformation.

Like ring chromosomes, inversions require two breaks in a chromosome. Paracentric inversions, that is, those which occur in one arm of a chromosome, cannot be identified in the mitotic chromosomes. Pericentric inversions, which may occur when the two breaks are on opposite sides of the centromere, can be recognized if the arm ratio of the chromosome is significantly changed. For example, one case of pericentric inversion involving chromosome 10 produced the change diagrammed in Fig. 2.17. A mother who was herself normal had one chromosome 10 looking like that shown in Fig. 2.17a. She had had several children with complex congenital malformations. One of these malformed children was found to have an ostensibly normal karyotype. It is likely, however, that one chromosome 10 represented a duplication-deficiency resulting from a crossover within the pericentric inversion, as diagrammed in Fig. 2.17b. Crossing over within the pericentric inversion can result in a great variety of recombinant chromosomes, each with duplication-deficiency. Thus, a person carrying an inversion can have several children afflicted with quite different types of malformations.

Abnormalities such as ring chromosomes, dicentrics, acentric fragments, and chromatid breaks have been observed in persons exposed to ionizing radiation or to radiomimetic agents, such as nitrogen mustard, that are used in treating some forms of cancer.

Chromosome Changes in Abortions

Ten to fifteen percent of all pregnancies terminate in spontaneous abortion before 20 weeks of gestation. Studies of chromosomes in aborted material confirm the suspicions that many have a chromosomal abnormality. The approximate frequency of specific abnormalities is shown in Table 2.3, column *A*. The relatively high frequency of the Turner syndrome (mainly with XO karyotype) and of Down's syndrome is of interest. Also note the rather frequent occurrence of

TABLE 2.3.   *Frequency of Selected Chromosomal Aberrations*

	A *Spontaneous* *abortions*	*B* *Live-born*
Sex chromosome abnormalities		
Turner syndrome (all types)	1/18	1/3,500 "females"
Klinefelter syndrome (all types)	Not found	1/500 "males"
Extra X chromosomes (mainly XXX)	Not found	1/1,400 "females"
Autosomal abnormalities		
Trisomy G	1/40	1/600
Trisomy 18	1/200	1/4,500
Trisomy D	1/33	1/14,500
Trisomy 16	1/33	almost 0
Triploidy	1/22	almost 0

karyotypes such as triploidy and trisomy 16, which are rarely observed in liveborn infants. Presumably trisomy of some other chromosomes is lethal at such an early stage that pregnancy is not even recognized as existing.

Chromosomal Changes with Aging

It has been found that older persons have an increasing number of cells with a chromosome number of 45. More specifically it has been found that the missing chromosome in females is an X chromosome and the missing chromosome in males a Y chromosome. In continually dividing cells chromosome aberrations, including loss of chromosomes, would be expected with aging. Loss of one X in the female or the Y in the male would not be expected to be lethal. The occurrence of these cells may be a reflection of a much larger amount of aberration from which cells do not survive. The possible relevance of the finding to furthering our understanding of the aging process is evident.

Chromosome Changes in Cancer

In many cases of a major form of leukemia, so-called chronic myeloid (or granulocytic) leukemia, a chromosomal aberration in the form of a deletion of part of the long arm of one of the four small acrocentric chromosomes (either 21 or 22), has been found. The abnormality is confined to blood cells as studied in cultures of peripheral blood or bone marrow and is not found in other somatic cells, for example, those derived from skin.

In leukemia as in Down's syndrome it is uncertain whether chromosome 21 or 22 is implicated. However, several observations have suggested that the same chromosome is involved, in different ways, in both Down's syndrome and leukemia: peculiarities in the shape of the nucleus of leukocytes occur in Down's syndrome. Leukemia in Down's syndrome patients is much more frequent than in the general population. Leukocyte alkaline phosphatase is higher than normal in these patients and lower than normal in those with myeloid leukemia.

The deletion of chromosome 21 in the leukocytes line, the "cause" of many cases of chronic myeloid leukemia, is presumably acquired, not inherited, as in the other anomalies discussed earlier. Causes of the deletion are under study; exposure to X rays may be one cause.

Ordinarily Mendelian disorders, that is, those which show simple patterns of inheritance as described in Chap. 3, do not show microscopically discernible abnormalities in the chromosomes. The genetic change, or mutation, in these cases is in DNA at a level far below the resolution of the microscope. Several conditions represent exceptions, however. For example, two conditions, called Fanconi's anemia and Bloom's disease after the men who described them, are inherited as autosomal recessive conditions, but patients suffering from them tend to show multiple chromosome breaks. Presumably the biochemical defect in each of these conditions is such that either the chromosomes are rendered more vulnerable to the effects of chromosome-breaking factors in the environment, such as viruses, or the mechanisms normally responsible for chromosome repair are faulty. Malignancy, especially leukemia, occurs commonly in patients with Fanconi's anemia or Bloom's disease, and may be a result of chromosomal breakage.

In summary, the chromosomal basis of certain congenital malformations, sex anomalies, behavioral abnormalities, spontaneous abortions, and neoplastic diseases has been elucidated. Chromosomal changes with aging have also been discovered. Chromosomal breakage leading to structural abnormalities, such as deletions, translocations, isochromosomes, and inversions, have been identified. As techniques improve, human cytogenetics will be able to detect more subtle changes and some may prove to be more important to man than the obvious changes already discovered.

Cytogenetic Mapping of the Chromosomes of Man

Mapping the chromosomes of man proceeds mainly by laboriously collecting linkage data on families (Chap. 3). However, cytogenetic approaches can help and can locate to specific chromosomes and parts of chromosomes the linkage groups established by family studies. General approaches that may be productive include the following:

(1) Deletion of part of one chromosome may "uncover" a recessive allele on the other chromosome and thus show that the locus is on the chromosomal segment which is missing. Patients with the *cri du chat* syndrome due to partial deletion of the short arm of one chromosome 5 should be scrutinized for the presence of a rare recessive disorder, since such would suggest that the locus of the gene for that disorder is on the deleted segment of chromosome 5. The conclusion is strengthened if appropriate tests show that only one of the parents is a carrier of the recessive gene. Polymorphic traits, such as blood groups, can also be used in this method of "deletion mapping" of the chromosomes.

(2) Familial distribution of marker traits may identify a particular locus with a particular chromosome. The following is a fictitious example: If the father of a trisomic offspring is blood type O, and if the mother and the offspring are both blood type AB, then the ABO locus must be on the chromosome involved in the trisomy, since the mother gave both alleles to the offspring. (See Chap. 4 for another possible explanation for such a finding, the Bombay gene.) Atypical frequencies of blood types in a trisomic population would be expected if the locus for that blood-group system is on the involved chromosome. However, it would be necessary to study a large number of subjects to demonstrate the effect.

(3) Dosage effects may point to the location of a particular gene on a particular chromosome. Leukocyte alkaline phosphatase activity is abnormally high in Down's syndrome (trisomy 21) and is abnormally low in cases of chronic myeloid leukemia with partial deletion of the long arm of chromosome 21. These findings suggest the presence of a "leukocyte alkaline phosphatase locus" on the long arm of chromosome 21. It is not certain, however, that excesses of certain enzymes in cases of Down's syndrome can be taken as evidence that the genes for those enzymes are on chromosome 21. In addition to leukocyte alkaline phosphatase, galactose-1-phosphate uridyl transferase, the enzyme whose deficiency occurs in galactosemia, has been found to be increased in Down's syndrome. However, the enzyme glucose-6-phosphate dehydrogenase (G6PD), whose structure is determined by a gene which is on the X chromosome, also is increased in Down's syndrome. Thus, the increase in activity of alkaline phosphatase and galactose-1-phosphate uridyl transferase in Down's syndrome may be due to some mechanism other than the location of a structural locus on the triplicated chromosome.

(4) Chromosomal peculiarities, such as unusually large satellites and the 15–21 translocation chromosome, can be used as one trait for family studies of linkage with other marker traits (Chap. 3). If a

giant satellite on chromosome 14 showed close linkage with a particular blood-group system, then one would conclude that that blood-group locus is on chromosome 14. Judging from data collected in families in which one of the chromosomes 1 is anomalous (Fig. 2.4), the locus for the Duffy blood groups may be on that chromosome (see p. 73).

(5) Somatic-cell genetics is in its infancy. The culture of human cells of known genotype may give mapping information. For example, after inducing chromosomal aberrations and separating pure sublines, one may be able to study the biochemical characteristics of each and make inferences as to the location of genes controlling those characteristics. Localization of the gene for thymidine kinase has been accomplished by somatic cell genetics. Mouse and human cells can be induced to fuse, when mixed in culture, especially if a particular virus is added. The experiment performed by Mary Weiss and Howard Green consisted of fusing a mouse cell lacking the capacity to synthesize thymidine kinase with a human cell which has this enzyme. The hybrid cell had normal thymidine kinase activity. With the passage of cell generations the human chromosomes were gradually lost from the hybrid. Finally, these workers ended up with a cell line which had only a single human chromosome remaining. Since the cell had thymidine kinase activity, the gene for this enzyme is thought to be located on the one remaining chromosome.

As will become evident in the chapters that follow, genetic analysis in man, as compared with many other species, is hampered by the inability to do experimental matings and by the long generation time. Genetic analysis by study of the biochemical characteristics of human cells in tissue culture may be a partial substitute for experimental matings. Transformation, mutation, visible chromosomal changes (induced or spontaneous) and hybridization through cell fusion are only a few of the approaches that are being explored, already with some success, in human tissue culture.

The Human Chromosomes in Meiosis

Up to this point, discussion has been confined largely to the chromosomes of somatic cells undergoing mitosis. However, a consideration of the chromosomes in meiosis, as well as other details of gametogenesis, is important, because chromosomal aberrations occur particularly in meiosis and point mutations of most genetic significance occur in the germ cells. For details of the mechanics of both mitosis and meiosis, see Swanson, Merz, and Young's *Cytogenetics,* in this series.

The reduction division which occurs at the first stage of meiosis reduces the diploid centromere number (46) to the haploid number (23) present in each gamete. At the beginning of meiosis, which occurs only in the germ cells, each chromosome is duplicated, and the now twin-strand homologous pairs undergo synapsis, that is, they come to lie side-by-side, attached at their centromeres. One member of each pair goes to each pole and the cell divides. This passage of one of a pair of homologous chromosomes into one daughter cell and the other into a second cell is the basis for Mendel's first law, that of the *segregation of alleles*. In the case of one pair of chromosomes, the maternally derived chromosome (or at least the chromosome with the maternally derived centromere) may pass into cell 1 and the paternally derived chromosome of the pair may pass into cell 2. Since it is an independent matter whether the paternal or maternal chromosome of another pair passes into cell 1 or cell 2, the events of the first stage of meiosis also constitute the basis of Mendel's second law, that genes on separate chromosomes assort independently.

In the latter part of meiosis, the daughter cells, each of which contains half the number of chromosomes (that is, half the number of centromeres) found in the original gonial cell, undergo a so-called equational division. As a result, gametes with half the somatic number of chromosomes are formed from these two divisions of the original germ cell.

When homologous chromosomes pair during the first stage of meiosis, they may exchange homologous segments, a process known as *crossing over*. An important consequence of meiosis is separation of the products of crossing over into different gametes. Through crossing over, the genetic diversity of the gametes is further increased; genes at different loci (nonalleles), even though on the same chromosome, may show assortment. If far apart on the chromosome, the assortment may be completely independent, as though the genes were on separate chromosome pairs. If the genes are closer together, less than completely independent assortment occurs. Studies of *genetic linkage* (p. 66) quantitate the degree of independence of assortment and thereby provide a means of mapping genes on the chromosome.

Human meiosis can be studied in testicular material (obtained after death or by biopsy performed for medical reasons) and prepared for examination by methods similar to those used for the mitotic chromosomes. Ovarian material is now also being studied by special techniques. The chromosomes in cells at the first meiotic metaphase show an association of homologous chromosomes in pairs called bivalents (Figs. 2.18 and 2.19). Study of this material provides at least four items of useful information.

(1) The X and Y chromosomes either are not joined or show terminal association (Fig. 2.18). Side-to-side pairing as shown by other chromosomes does not occur. This cytologic discovery suggests that the X and Y chromosomes of man have little or no homologous segment and that crossing over between them does not occur. If any homologous segment did exist, genes carried on it would produce traits with a

characteristic pattern of transmission in families, so-called partial sex linkage. There is, however, no substantial evidence from family studies that any trait in man is inherited in this manner.

(2) Chiasmata between homologous autosomes are visualized in meiotic material and can be counted. The formation of chiasmata may represent the physical exchange of chromosomal material between homologous chromosomes; in any event, chiasmata are correlated with genetic crossing over. *Chiasma* literally means "cross" and is the physical basis of the genetic crossing over that can be shown in family studies. The total number of chiasmata is a measure of the genetic length of the genome; the unit of map distance in genetic linkage studies is the crossover unit, or recombination fraction. Chiasma counts indicate that the total genetic length of the human chromosomes is approximately 3,000 map units (p. 72).

(3) Chromosomal rearrangements such as translocations and inversions are most easily recognized in meiotic material from the configuration of the bivalents. For example, characteristic translocation figures are found in the male carrier of the 15–21 translocation chromosome.

FIG. 2.18. *The human chromosomes in meiosis in the male. The X and Y chromosomes are attached terminally* (dotted line). *Courtesy of M. A. Ferguson-Smith.*

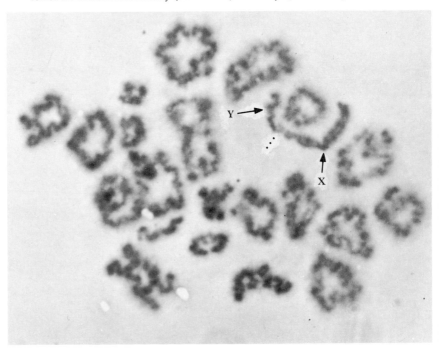

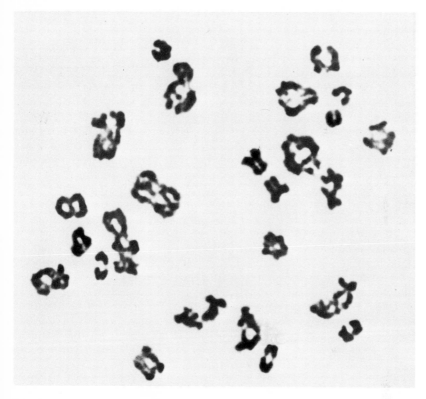

FIG. 2.19. *The human chromosomes in meiosis in the female; metaphase of the first stage of meiosis. The two X chromosomes cannot be identified with certainty. Courtesy of Georgiana Jagiello.*

(4) In man, as in many other species, the chromosomes in the pachytene stage of prophase show chromomeres—banding with characteristic spacing and size. Only beginnings of a pachytene map in man have been made, and the goal of a complete pachytene map, correlated with a genetic map derived from linkage and cytogenetic studies, is yet to be achieved.

Oogenesis and Spermatogenesis in Humans

Human female germ cells multiply rapidly during early fetal life. Oogonia (the primordial germ cells of the female) cease to propagate after the fifth or sixth month of fetal life. The female infant is born with a full stock of oocytes that must last for her entire reproductive life. Inventories of this stock have arrived at estimates of about

750,000 per individual female; however, a large proportion of these degenerate at various stages of oogenesis and at various times in the life of the female.

By about the time of birth the oocytes have completed most of the prophase of the first meiotic division and then regress into a long interphase-like dictyotene stage during which the nuclear membrane remains intact and the chromosomes are visible as thread-like or net-like structures. The dictyotene stage lasts for at least 12 years and even as long as 50 years! During this period the DNA content of the oocyte is tetraploid. (The DNA content of diploid interphase somatic cell nuclei is about 5.5×10^{-12} gm per cell. Tetraploid liver cells and oocytes in the long dictyotene stage have double this amount.)

The first meiotic, or reduction, division of the oocyte is not completed until about the time of ovulation. Through the influence of pituitary hormone the dictyotene stage is terminated and resumption of meiosis is induced. The second, or equational, division of meiosis is usually completed only after the entry of sperm into the ovum. A polar body as one product of each meiotic division is extruded and lost. The second meiotic division separates the products of the crossing over that occurred between chromatids during prophase of the first meiotic division. When the chromosome groups in the male and female pronuclei come together, fertilization is complete and the cell becomes a zygote. The process of fertilization usually takes place in the ampulla of the Fallopian tube. Implantation of the young human embryo occurs six to seven days after fertilization.

The fact that the second meiotic division of the egg does not occur until after entry of the sperm is probably related to the development of triploid individuals (Table 2.3) who have a $3n$ chromosome complement. Either dispermy or fertilization of a diploid egg can occur. Also, the development of mosaics by double fertilization (p. 26) occurs at this critical stage.

Spermatogenesis differs from oogenesis in several important respects. The production of sperm is exceedingly abundant and the total number produced in the lifetime of a male is truly astronomical. Proliferation of spermatogonia does not begin until puberty, but thereafter usually continues throughout the lifetime of the male; oogenesis, on the other hand, is confined to the intrauterine life of the female. Four spermatids rather than one are produced from each spermatogonium; no polar bodies are produced. The time for completion of the full cycle of spermatogenesis is about 64 days, and not from 12 to 50 years as in oogenesis. The two X chromosomes of the oogonia are isopycnotic, that is, they stain identically and stain like the autosomes, and by genetic and cytologic data they show crossing over between them. On the other hand, the X and Y chromosomes of the spermatogonium are heteropycnotic. They undergo early condensation and show at the most terminal association (Fig. 2.18). This mechanism has probably evolved to guarantee isolation of the critical male-determining genetic material.

The extent to which the genetic constitution of sperm is expressed in the phenotype of the sperm is a question of great biological and practical importance. For example, does the sperm carrying the allele for blood group A show this antigenetic specificity? The answer to this question is important to the matter of gametic selection in maintenance of the ABO polymorphism. Is there reduced fertility in couples of which one is homozygous for blood group A and the other homozygous for blood group B, and if so, might the reduced fertility be explained by the relative suppression of A-bearing sperm by anti-A antibody of the mother, or vice versa, of B-bearing sperm by anti-B of the mother? Among the sperm of AB men, are the A sperm at a relative disadvantage in women of blood type B, and B sperm at a disadvantage in women of blood type A? Full answers to these questions are not yet available. Selection may be operating on sperm, but the details are as yet largely unknown.

The ability to separate classes of sperm according to haploid genotype could have great practical importance. Control of the sex of the offspring is only one potential use. If one could separate into two classes the sperm from a male heterozygous for a gene causing a grave disease, one could perhaps avoid the transmission of that gene to the next generation—to the benefit both of his family and of the population in general. Unfortunately, however, even the dimorphism that should be most striking—that between X-bearing and Y-bearing sperm —is not reliably discernible by any method yet devised.

Mutation: Change in Chromosomes

In the general sense, genetic mutation includes both (1) chromosomal aberrations such as those detectable by presently available, relatively gross techniques and (2) point mutations. Point mutations, which are detected by the presence of simply inherited disorders or traits in subsequent generations, occur in the germ cells of either the male or the female. Mutation also occurs in somatic cells, where it does not differ in principle from that in the germ cells. If somatic mutation occurs in the zygote or at an early post zygotic stage, the individual is affected with the given disorder if it is a dominant. Furthermore, if a somatic mutation occurs at an appropriate postzygotic stage early in embryogenesis and if the action of the mutant gene is locally manifest, a sectorial, or mosaic, situation results. For example, if the normal allele corresponding to the allele for von Recklinghausen's neurofibromatosis underwent mutation in a somatic cell early in embryogenesis, one might observe a sector of skin in which the characteristic tumors and pigmented spots occur. Such an individual would not transmit the disorder unless gonadal tissue also was involved. When a gonadal anlage cell suffers mutation, the situation is referred to as gonadal (or germinal) mosaicism. Such an individual, although not showing signs of the disorder even though it is dominant, may have two or more affected children. Gonadal mosaicism for chromosomal aberrations has been observed. Normal women who have had two children with Down's syndrome have been found to have a minor trisomy-21 cell line among the circulating lymphocytes and/or skin fibroblasts. Presumably part or all the oogonia are trisomic for chromosome 21.

Each spermatogonium produces many mature sperm. Point mutation occurring in a spermatogonium can, therefore, result in many sperm which carry the mutation. Persistence of the spermatogonial cell carrying the mutation results in what amounts to gonadal mosaicism, even though only one cell out of many is different. Spermatogonial mutation is presumed to be the basis for the "paternal age effect" demonstrable for a number of autosomal dominant conditions (see p. 51). It may be the basis for the occasional observation of two or more sibs affected by a dominant condition which is present in neither parent.

A third important class of mutation is that resulting from unequal crossing over. Normally, during meiosis, homologous chromatids break and rejoin with exchange of segments, so-called crossing over. If non-homologous pairing and unequal crossing over occur, deletion of a segment of chromosome or duplication of a segment results. The situation diagrammed in Fig. 2.20a is that when the break occurs between genes. One of the products has a duplication of gene B. Another has no gene B—possibly a lethal situation. Evolution is thought to have occurred through gene duplication of this type. Independent mutation can occur in the second locus with exploration of possible adaptive advantages, without losing the benefit of the previously existing gene. The gene determining the gamma chain of fetal hemoglobin (see p. 95) has been duplicated and mutation has occurred in one of the daughter genes so that two types of gamma chain differing by one amino acid are present in each person. The several types of hemoglobin genes (alpha, beta, delta, and gamma) appear to have evolved from a common primordial gene through gene duplication and separate mutation (p. 176).

How deductions about gene duplication are possible on the basis of analysis of the amino acid sequence of product proteins is discussed in Chap. 4. An example of deletion through the mechanism of unequal crossing over is hemoglobin Lepore (p. 74), which has an abnormal polypeptide chain made up of part of the beta chain and part of the delta chain. Figure 2.20b shows the manner in which this mechanism appears to function. In certain other abnormal hemoglobins whose amino acid sequence has been determined, deletion of one amino acid or several sequential amino acids has taken place, presumably by this

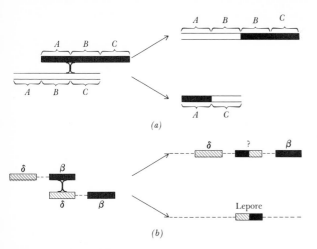

FIG. 2.20. *Nonhomologous pairing and unequal crossing over: (a) when the break occurs between genes; (b) when the break occurs within a gene, as appears to have occurred to produce the anomalous polypeptide chain in hemoglobin Lepore.*

mechanism of unequal crossing over. This third class of mutation can, if you prefer, be viewed as a special form of the second class, chromosomal aberrations.

References

Carr, D. H., "Chromosome Studies in Spontaneous Abortions," *Obstetrics-Gynecology, 26* (1965), 308–26.

Clermont, Y., "Renewal of Spermatogonia in Man," *Am. J. Anat., 118* (1966), 509–24.

Court Brown, W. M., *Human Population Cytogenetics.* Amsterdam: North-Holland Publishing Co., 1967.

De Capoa, A., D. Warburton, W. R. Breg, D. A. Miller, and O. J. Miller, "Translocation Heterozygosis: A Cause of Five Cases of the *Cri du Chat* Syndrome and Two Cases with a Duplication of Chromosome Number Five in Three Families," *Am. J. Human Genet., 19* (1967), 586–603.

Donahue, R. P., W. B. Bias, J. H. Renwick, and V. A. McKusick, "Probable Assignment of the Duffy Blood Group Locus to Chromosome 1 in Man," *Proc. Natl. Acad. Sci. U.S., 61* (1968), 949–55.

Edwards, J. H., Catherine Yuncken, D. I. Rushton, Susan Richards and Ursula Mittwoch, "Three Cases of Triploidy in Man," *Cytogenetics, 6* (1967), 81–104.

Ferguson-Smith, M. A., "Chromosomes and Human Disease," in *Progress in Medical Genetics*, Vol. I, Arthur G. Steinberg, ed. New York: Grune & Stratton, Inc., 1961.

Hamerton, J. L., "Robertsonian Translocations in Man: Evidence for Pre-zygotic Selection," *Cytogenetics, 7* (1968), 260–76.

Jagiello, G., J. Karnicki, and R. J. Ryan, "Superovulation with Pituitary Gonadotrophins: Methods for Obtaining Meiotic Metaphase Figures in Human Ova," *Lancet, 1* (1968), 178–80.

McKusick, Victor A., "On the X Chromosome of Man," *Quart. Rev. Biol., 37* (1962), 69–175; also American Institute of Biological Sciences Monograph, 1964.

Montagu, M. F. A., ed., *Genetic Mechanisms in Human Disease: Chromosomal Aberrations.* Springfield, Ill.: Charles C Thomas, Publisher, 1961.

Ohno, S., H. P. Klinger and N. B. Atkins, "Human Oogenesis," *Cytogenetics, 1* (1962), 42–51.

Smith, D. W., J. M. Docter, P. E. Ferrier, J. L. Frias, and A. Spock, "Possible Localisation of the Gene for Cystic Fibrosis of the Pancreas to the Short Arm of Chromosome 5," *Lancet, 2* (1968), 309–12. But see Danes, B. S., and A. G. Bearn, "Localisation of the Cystic-fibrosis Gene," *Lancet, 2* (1968), 1303.

Swanson, Carl P., Timothy Merz, and William J. Young, *Cytogenetics.* Englewood Cliffs, N.J., Prentice-Hall, Inc., 1967.

Weiss, M. C., and Howard Green, "Human-Mouse Hybrid Cell Lines Containing Partial Complements of Human Chromosomes and Functioning Human Genes," *Proc. Natl. Acad. Sci., 58* (1967), 1104–11.

Wilson, Edmund B., *The Cell in Development and Heredity.* New York: The Macmillan Company, 1925. A classic of biology.

Yerganian, George, "Cytologic Maps of Some Isolated Human Pachytene Chromosomes," *Amer. J. Human Genet., 9* (1957), 42–54.

Zuelzer, W. W., K. M. Beattie, and L. E. Reisman, "Generalized Unbalanced Mosaicism Attributable to Dispermy and Probable Fertilization of a Polar Body," *Am. J. Human Genet., 16* (1964), 38–51.

Supplement to Chapter 2:
A Shorthand for Describing the Karyotype

Recommendations for standardized ways to represent the karyotype briefly in publications were made by a group which met in Chicago in 1966 (published by the National Foundation—March of Dimes in the *Birth Defects Original Article Series,* Vol. II, No. 2). Examples following these recommendations are given below. Note that *p* (for *petite*) and *q* are the designations for the short and long arms, respectively. A diagonal line is used to separate cell lines in describing mosaicism. A plus sign (+) or a minus sign (−) after an autosome number or group letter indicates that the particular chromosome is extra or missing; when used immediately after an arm designation (*p* or *q*), it means that that part is abnormally long or abnormally short. Satellite is referred to by *s,* translocation by *t,* isochromosome by *i.*

45,X	Usual karyotype of Turner syndrome.
47,XXY	Usual karyotype of Klinefelter syndrome.

47,XX,G+	Karyotype of female with Down's syndrome.
45,X/46,XX/47,XXX	A triple cell line mosaic.
46,XY,5p—	Karyotype of male with *cri du chat* syndrome.
46,XY,Gq—	Karyotype of leukocytes of male with myeloid leukemia.
46,XY,t (Bp—;Dq+)	Karyotype of male with balanced reciprocal translocation between short arm of a B and long arm of a D group chromosome.
45,XX,D—,G—,t (DqGq) +	Karyotype of female translocation Mongolism carrier.
46,XY,14s+	Karyotype of male with giant satellited chromosome 14.
46,XXqi	Karyotype of female with Turner syndrome due to isochromosome of the long arm of one X chromosome.

Three

Genes in Kindreds

The chief method of genetic study in man is the observation of pedigree patterns, that is, the patterns of distribution of genetic traits in kindreds. Since critically informative matings cannot be made by design, as is possible in experimental genetics, the human geneticist must rely on collections of families for information on the genetics of a given trait. The pedigree pattern provides information on the Mendelian principles of segregation and independent assortment; furthermore, it may provide information on allelism and linkage. Single-factor inheritance is studied by pedigree patterns. Many important traits, such as intelligence, are, however, determined by many collaborating genes. Here, too, analysis of intrafamilial similarities, essentially an extension of the pedigree method, is the principal approach.

Study of a particular trait in a family usually begins with an "affected" person who is referred to as the *proband*, or the *propositus* (female = *proposita*), or (especially by epidemiologists) the *index case*. The following conventions are useful in the construction of pedigree charts. The use of squares for males and circles for females predominates in this country and in continental Europe, whereas the symbols of Mars and Venus ($\male$ and $\female$) are more generally used in Britain. Breeding records of species other than humans generally list the female first in the representation of matings, that is, $\female \times \male$. In human pedigree charts the usual practice is to place the male first, on the left,

that is, ⬜—○. Some prefer to set down the marital line thus: ⌐ . Some prefer to set down the marital line thus: ⌐.

Double marital lines are used in cases of consanguinity: ⬜═○. The proband is marked by an arrow.

Sibs are indicated thus: ○₁ ⬜₂ ○₃ ○₄ in chronological order

of birth. For economy of space, the number of normal sibs can be

shown as: ⬜₃ ○₂ or ⬜₅ or ○₅ . Abortions or stillbirths are indicated

by small symbols: ● ; recording the sex, when known, as M or F, is

valuable. Twins are indicated by: ⬜—⬜ or ○—○ if monozygotic;

by ⬜ ⬜ , ○ ○ , or ⬜ ○ if definitely dizygotic; and by ⬜ ? ⬜

or ○ ? ○ if of uncertain zygosity.

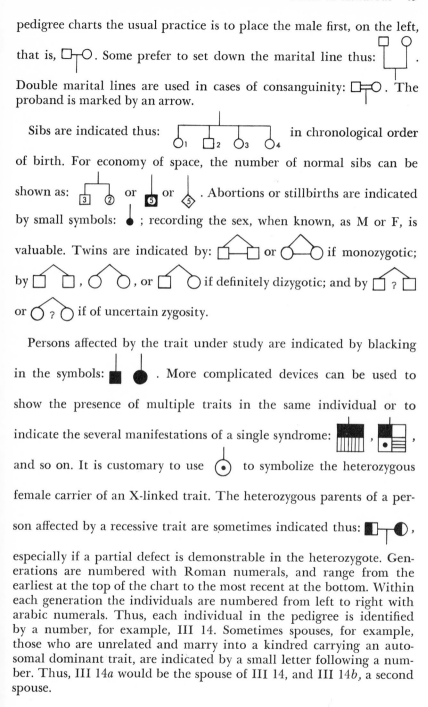

Persons affected by the trait under study are indicated by blacking

in the symbols: ■ ● . More complicated devices can be used to

show the presence of multiple traits in the same individual or to

indicate the several manifestations of a single syndrome: ▦ , ▦ ,

and so on. It is customary to use ⊙ to symbolize the heterozygous

female carrier of an X-linked trait. The heterozygous parents of a per-

son affected by a recessive trait are sometimes indicated thus: ◨—◖ ,

especially if a partial defect is demonstrable in the heterozygote. Generations are numbered with Roman numerals, and range from the
earliest at the top of the chart to the most recent at the bottom. Within
each generation the individuals are numbered from left to right with
arabic numerals. Thus, each individual in the pedigree is identified
by a number, for example, III 14. Sometimes spouses, for example,
those who are unrelated and marry into a kindred carrying an autosomal dominant trait, are indicated by a small letter following a number. Thus, III 14a would be the spouse of III 14, and III 14b, a second
spouse.

Mendel's Principle of Segregation

In 1866 G. Mendel wrote, "Henceforth . . . those characters which are transmitted entire, or almost unchanged in the hybridization, and therefore in themselves constitute the characters of the hybrid, are termed the *dominant,* and those which become latent in the process *recessive"* (italics mine). The definition is precisely the one currently used if the word *heterozygote* is substituted for *hybrid.* Dominant traits are those that are expressed in the heterozygote. Recessive traits[1] are expressed only in the homozygote. It is implicit in the definition that the terms *dominant* and *recessive* refer, strictly speaking, to characters, not to genes. However, for convenience geneticists often speak, especially in population genetics, of a "recessive gene" or a "dominant gene."

The part of Mendel's definition, "almost unchanged in the hybridization," allows for phenotypic differences in the dominant character in the homozygote as compared to the heterozygote. The reader will recall that Mendel's classic experiments involved crosses between two. homozygotes, one with the dominant character and one with the recessive character.

Many rare traits in man are distributed in families in characteristic patterns that are in accordance with the laws of Mendel. The specific pedigree pattern is dependent on whether the responsible mutant gene is located on one of the autosomal chromosomes or on an X chromosome, and also on whether the effects of the gene are evident in single dosage, that is, in the heterozygous state, or whether the gene must be present in double dosage (or the homozygous state) for expression. Depending on the type of chromosome bearing the gene in question, a trait is said to be either autosomal or X-linked. (The older term *sex-linked* for X-linked is less satisfactory for several reasons, among them the fact that, if they exist, holandric Y-linked traits are also "sex-linked.") Depending on whether expression of the gene occurs in the heterozygous state or only in the homozygous state, a trait is said to be dominant or recessive, respectively.

Autosomal Dominant Inheritance

The first pedigree to be interpreted in terms of Mendelian dominant inheritance was studied by Farabee when he was a graduate

---

1 As used by the geneticist, "trait" refers to any phenotype, whether disease or "normal variation," and whether in the homozygote (as in a recessive condition) or heterozygote (as in a dominant condition). In clinical medicine "trait" came to be applied to the heterozygous state of the sickle hemoglobin gene and by analogy is often used for the heterozygous state of other recessive genes. In the discussion here, "trait" will be used in the general phenotypic sense of the geneticist.

student at Harvard in the early years of this century. In the family he studied, the trait brachydactyly (short fingers) can be seen to be segregating (Fig. 3.1). Since it is rare, it usually occurs in one parent only. The unaffected spouses are omitted from this chart. On the average, half the sons and half the daughters of a man or a woman with brachydactyly are affected. This result is to be expected when the responsible gene is located on one of a pair of autosomes, since only one of each pair is contributed to a given offspring by the affected parent. The chromosome may with equal likelihood be either the one bearing the mutant gene for brachydactyly or the one bearing its normal, or so-called wild-type, allele.

In general, dominant traits are less severe than recessive traits. In part, an evolutionary or selective reason for this observation can be offered. A dominant lethal,[1] that is, a dominant mutation that determines a grave disorder making reproduction impossible, will promptly disappear because it will not be transmitted to the next generation. On the other hand, a recessive mutation, even if in the homozygous condition it precludes reproduction, can gain wide dissemination in heterozygous carriers.

Among children whose parents are both affected by a dominantly inherited trait, i.e., are both heterozygous for the responsible gene, the genotypic expectations are precisely as described by Mendel: one-fourth are homozygous affected, one-half heterozygous, and also affected, and one-fourth homozygous normal. Few autosomal dominant traits have ever been observed in the homozygous state. Some disorders that are relatively mild in the heterozygous state are lethal in the homozygous state; see Fig. 3.2 for a probable example. Probably few traits are completely dominant, that is, have the same phenotypic expression when the gene is in the heterozygous state as when it is in the homozygous state.

Another characteristic of dominant traits is wide variability in severity, or "expressivity" (see Chap. 4). Sometimes the expressivity is reduced so much that the gene cannot be detected, at least by the methods currently available. When this is the case, the trait is said to be "nonpenetrant." So-called skipped generations sometimes occur in pedigrees of families with a dominant trait. In the skipped individual expressivity is so low that the presence of the gene is not recognizable from the phenotype, that is, the trait is nonpenetrant in that person. Sometimes in taking a family history an apparent skipped generation turns up. However, when the skipped individual is subjected to close study, he in fact shows definite although mild manifestations.

Another phenomenon based largely on the wide variability in the severity of dominant traits is so-called anticipation: the given hereditary disease manifests itself earlier, is more severe, and leads to earlier

[1] *Lethal* as used here refers to "genetic death." If a mutant gene has the effect that its bearer cannot bear offspring even though he may survive to an advanced age, the genetic consequences are the same as if the gene led to death *in utero* (that is, abortion).

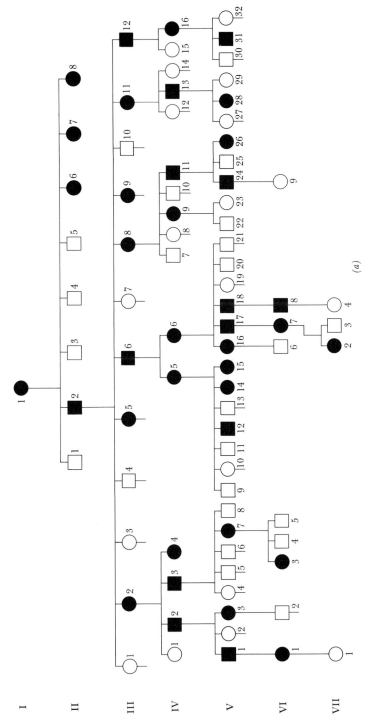

(a)

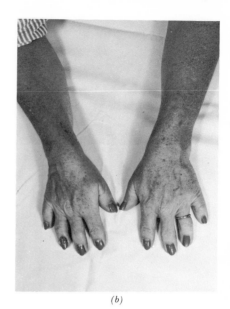

FIG. 3.1. (a) **Facing page.** *The family with brachydactyly first reported by Farabee in 1903; brought up-to-date in 1962.* (b) **Brachydactylous hands.**

(b)

death in each successive generation, or at least in one generation than in the one immediately preceding it. This has been purported for myotonic dystrophy, a form of muscular dystrophy. Obviously the affected persons in generation I that have children are those at the milder end of the bell-shaped normal distribution curve of severity, whereas *their* affected children, in generation II, will more nearly cover the whole range of severity. On the average, then, generation II is likely to be more severely affected than generation I by any measure, such as age at onset, degree of incapacitation, or age at death. Another source of bias leading to the artifactual phenomenon of anticipation is the fact that parent-offspring sets used in the calculations are often ascertained through the children. The more severely affected these children are, the more likely they are to come to attention. Anticipation has no genuine biological basis.

One cannot expect that all persons with a particular autosomal dominant trait will have one parent affected, let alone an extensive

FIG. 3.2. *Both parents had hereditary hemorrhagic telangiectasia of about average severity. Their child was severely and lethally affected with multiple angiomatous malformations in many organs. This may have been an example of lethal homozygosity for a rare autosomal dominant gene. Redrawn from L. H. Snyder C. A. Doan, J. Lab. Clin. Med., 29 (1944), 1211.*

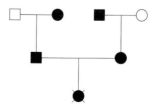

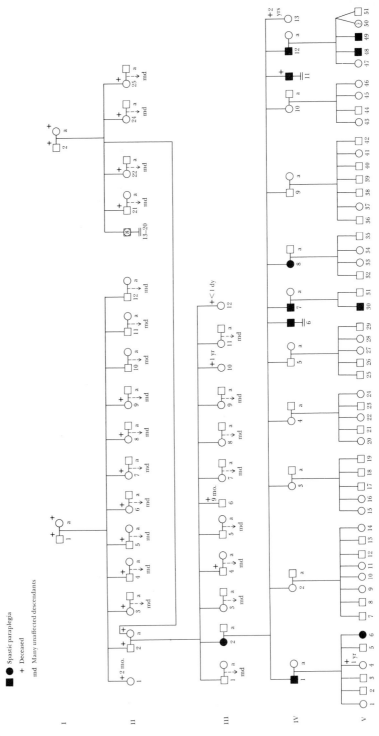

FIG. 3.3. *New dominant mutation. The condition is spastic weakness of the legs. The condition appeared de novo in a closed Amish community in which illegitimacy could be confidently excluded and no other case than those indicated here had occurred. Presumably fresh mutation occurred in a germ cell of the father or mother of person III 2. From V. A. McKusick and co-workers,* **Cold Spring Harbor Symp. Quant. Biol., 29 (1964), 99–114.**

pedigree such as that shown in Fig. 3.1. Unless there is mistaken paternity, phenocopy or a recessively inherited genetic mimic, or incomplete penetrance in one parent, the condition in the individual at hand may be sporadic (that is, nonfamilial), having resulted through the mechanism of fresh mutation in a germ cell of either the father or the mother (see Fig. 3.3). The more severely the condition interferes with reproduction of affected individuals, the larger is the proportion of cases which result from new mutation. This follows from the equilibrium which presumably exists between the addition of new genes to the gene pool through new mutation and the removal of genes from the gene pool through negative selection. The situation is analogous to a bucket in which water is entering from a tap and flowing out through a hole in the bottom: a situation of equilibrium can be attained at which inflow and outflow match each other and the level of water in the bucket remains constant. (These matters are discussed further in Chap. 7. See particularly Fig. 7.1 on p. 166 and the accompanying discussion on p. 165.) On the average, achondroplastic dwarfs reproduce at a frequency of only about one-eighth that of the general population. About seven-eighths of affected families have only one case, the proband, presumably the result of new mutation.

For several dominantly inherited conditions it has been possible to demonstrate paternal age effect. "Maternal age effect" is well known in connection with certain chromosomal aberrations, notably Down's syndrome (p. 21). Paternal age effect is demonstrable in the case of many fresh dominant mutations. Thus, if the age of fathers of sporadic cases of achondroplastic dwarfism is averaged, it is found to be 5 to 7 years higher than that of the general population of fathers.

It is theoretically possible for a dominant trait to show a pedigree pattern like that shown in Fig. 3.4. A mechanism by which a dominant trait might occur in two or more offspring of normal parents, with regular dominant transmission thereafter, is germinal mosaicism. If mutation occurs in the early embryo so that part or all of the gonad carries the mutation although none (or no significant part) of the body tissues are affected, then the pedigree pattern shown in Fig. 3.3 might occur. This pedigree pattern could also result from mutation in a single spermatogonium which was the progenitor of many sperm over a period of years. Also, the variability notorious in dominant traits might fortuitously result in such a pedigree pattern, the condition being "nonpenetrant" in the affected parent of the first generation. Of

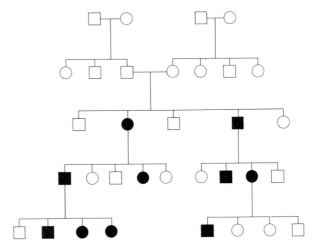

FIG. 3.4. *Pedigree pattern of autosomal dominant trait originating in a gonadal mosaic or spermatogonial mutation.*

course, illegitimacy (better, nonpaternity or mistaken paternity) could also account for this pedigree pattern if the true father in the first generation was affected. Although it is difficult to gauge its frequency, because it is hard to establish this mechanism in individual cases, germinal mosaicism almost certainly occurs. Several pedigrees of the lobster claw deformity, or ectrodactyly ("abortive fingers"), have been consistent with germinal mosaicism.

Autosomal Recessive Traits

Traits inherited as autosomal recessives (Fig. 3.5) likewise occur with equal frequency in males and females. When the trait is rare almost all the affected individuals have normal parents, but both are heterozygotes. Autosomal recessive inheritance is inheritance from both parents. Since related individuals are more likely to be heterozygous for the same mutant gene than are unrelated individuals, consanguineous matings, of first cousins for example, have a higher probability of producing offspring affected by a recessive trait. Viewed in another way, it is likely that a greater proportion of the parental matings in families affected by recessive traits are consanguineous than is true generally. The rarer the recessive trait, the higher the proportion of consanguineous parental matings. For more frequent autosomal recessive disorders (cystic fibrosis of the pancreas may be an example), there is little or no more consanguinity among the parents than expected by chance. In the case of a very rare recessive trait the occurrence of parental consanguinity may be the first clue to the fact that the trait is genetic.

Among the offspring of two heterozygous parents, one-fourth of males and females are expected to be homozygous and affected. However, in human genetics, sibships with both parents heterozygous are generally ascertained only through the occurrence in them of at least one affected member. Since there is usually no way to recognize the matings of two appropriately heterozygous parents who are so fortunate as to escape having affected children, a collection of sibships containing at least one affected child is a biased sample. In the *ascertained* families more than the expected one-fourth are affected. Bias of ascertainment is more fully discussed on p. 133.

If an individual affected by a recessive trait marries a homozygous normal person, none of the children will be affected, but all will be heterozygous carriers. If an individual affected by a recessive trait marries a heterozygous carrier of the same recessive gene, one-half of the offspring will, on the average, be affected, and a pedigree pattern superficially resembling that of a dominant trait will result. It was previously thought that two genetic forms of alkaptonuria existed— one inherited as an autosomal recessive and one as an autosomal dominant. Closer investigation revealed that the apparently dominant form was the same disease as the clearly recessive one. Because of in-

FIG. 3.5. *Pedigree pattern of a rare autosomal recessive trait, hereditary microcephaly. All four parents of affected persons are descendant from a couple who were born in the 1700s and at least one of whom was presumably heterozygous for the rare microcephaly gene. From V. A. McKusick and colleagues, "Chorioretinopathy with Hereditary Microcephaly,"* Arch. Ophthal., 75 (1966), 597–600.

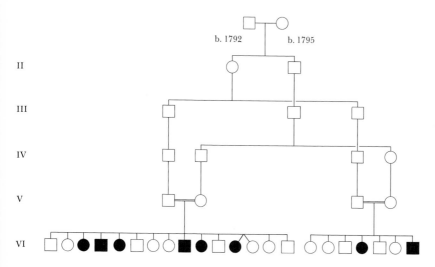

breeding, homozygous affected individuals frequently mated with heterozygous carriers, and a quasi-dominant pedigree pattern resulted (see Fig. 3.6) .

Dominance and recessiveness are somewhat arbitrary and artificial concepts. When our methods are sufficiently acute the effect of a recessive gene in the heterozygous state can often be recognized. Furthermore, a gene that has obvious expression in the heterozygous individual

FIG. 3.6. (a) *An incomplete pedigree showing quasi-dominant inheritance of alkaptonuria.* (b) *The complete pedigree showing that consanguinity accounts for the pedigree pattern simulating autosomal dominant inheritance. After A. K. Khachadurian and K. A. Feisal,* J. Chronic Diseases, 7 (1958), 455.

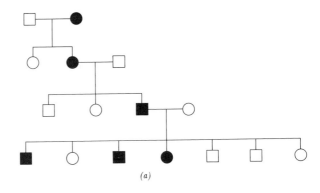

(a)

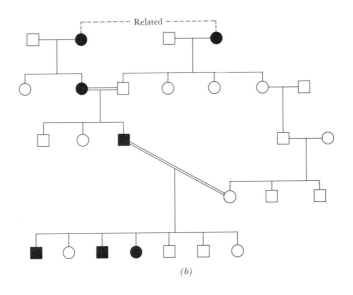

(b)

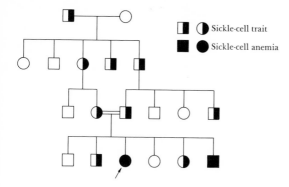

Sickle-cell trait

Sickle-cell anemia

FIG. 3.7. *Pedigree of a kindred with sickle-cell anemia showing that dominance and recessiveness are characteristics of the phenotype, not of the gene.*

and is therefore considered dominant may have a different effect, quantitatively and even qualitatively, in the homozygous state. The gene for sickle hemoglobin (Hb S) and the states referred to as sickle-cell anemia and sickle-cell trait illustrate the arbitrary nature of the distinction. The phenotype sickle-cell anemia is recessive since a homozygous state of the gene is required (Fig. 3.7). Sickling, however, is a dominant phenotype, since the gene in heterozygous state is expressed. *Intermediate inheritance* is the term sometimes applied to this type of pedigree pattern. Intermediate inheritance means that the heterozygous individual is identical to neither of the homozygous individuals but is in a sense intermediate. *Incompletely dominant* or *incompletely recessive* are yet other terms used for this intermediate situation.

Codominance is the term used for characters that are both expressed, or jointly expressed, in the heterozygote. For example, persons with the blood group AB demonstrate the effects of both the gene for antigen A and the gene for antigen B. Neither is recessive to the other. Similarly, the genes for different hemoglobins are both expressed if the method for demonstrating the phenotype is paper electrophoresis, such as in a person with both Hb S and Hb C. These examples of codominance again indicate that whether we view the phenotype as recessive or dominant is dependent largely on the acuteness of our methods for recognizing the products of gene action.

X-linked Inheritance

Like autosomal traits, those determined by genes on the X chromosome may be either dominant or recessive. The female with two X chromosomes may be either heterozygous or homozygous for a given mutant gene; the trait in the female can demonstrate either re-

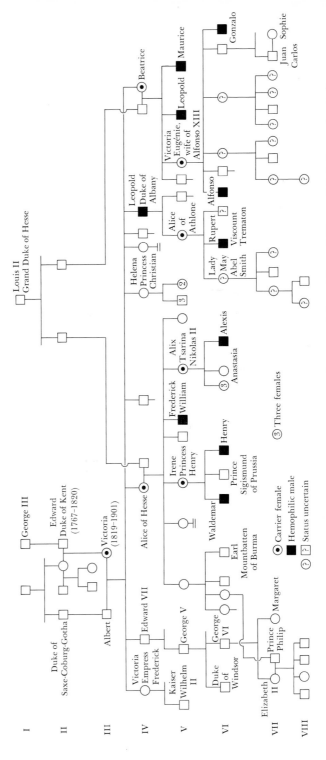

FIG. 3.8. *Pedigree of Queen Victoria and her descendants, illustrating the transmission of hemophilia, an X-linked recessive trait.*

56

cessive or dominant behavior. But the male with one X chromosome can have only one genetic constitution, namely hemizygous, and regardless of the behavior of the gene in the female, whether recessive or dominant, it is always expressed in the male.

The critical characteristic of X-linked inheritance, both dominant and recessive, is the absence of male-to-male, that is, father-to-son transmission. This is a necessary result of the fact that the X chromosome in the male is transmitted to none of his sons although it passes to each of his daughters.

X-linked recessive inheritance is illustrated in a classical manner by hemophilia. Queen Victoria was a carrier (Fig. 3.8). Since none of her forebears or collateral relatives were affected the mutation may have occurred in an X chromosome in the germ line in one of her parents or in her early embryonic stage. One of her sons, Leopold, Duke of Albany, died of hemophilia at the age of 31. Prince Albert, the consort of Victoria, can be exonerated since male-to-male inheritance is impossible. At least two of Victoria's daughters were carriers for hemophilia since several male descendants were hemophiliacs. In the Czarevitch, son of the last Czar of Russia, and in the princes of Spain the gene for hemophilia inherited from Victoria had considerable political consequences.

The pedigree pattern of an autosomal dominant trait tends to be a vertical one with the trait passed from generation to generation. The pedigree of an autosomal recessive trait tends to be horizontal, with affected persons confined to a single generation. The pedigree pattern of a rare X-linked recessive character tends to be oblique because the affection is almost exclusively of males and the transmission is to the sons of their normal, carrier sisters. William Bateson compared this pattern to the knight's move in chess. Tracing X-linked recessive characters through many generations is often difficult because the patronymic of affected persons usually changes with each generation.

Among the children of a male affected by an X-linked recessive trait, all sons are unaffected and all daughters are carriers (Fig. 3.8) —providing the mother is not affected and is not a heterozygous carrier. Father-to-son transmission of X-linked traits cannot occur. The father gives his one X chromosome to each daughter but to none of his sons.

To have hemophilia a female must be homozygous for this recessive gene; she must have received a gene for hemophilia from each parent. This can occur, and has been observed, when a hemophilic male marries a carrier female. As with other rare recessive traits this homozygous state is more likely to result from consanguineous matings (Fig. 3.9). Occasionally females with hemophilia are "manifesting heterozygotes."

In man one can enumerate about seventy traits, most of them pathologic, that are certainly X-linked. Most of them are X-linked recessives.

In X-linked dominant inheritance both males and females are af-

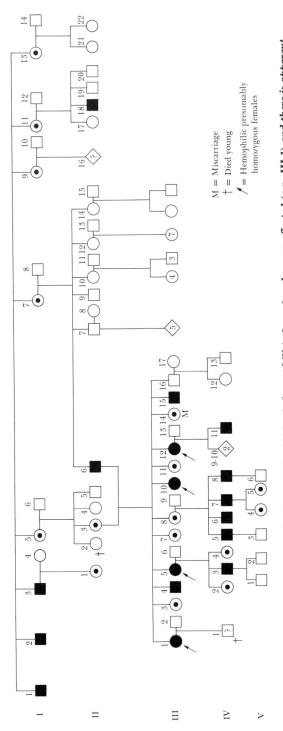

FIG. 3.9. The pedigree of a kindred with **hemophilia A** (classic hemophilia). Some females were affected (e.g., III 1) and there is apparent male-to-male transmission. However, affected males such as III 4 received the hemophilia gene from the carrier mother. Based on L. Gilchrist, Proc. Roy. Soc. Med., 54 (1961), 813.

58

fected and both males and females transmit the disorder to their offspring, just as in autosomal dominant inheritance. Superficially the pedigree patterns in the two types of inheritance are similar, but there is a critical difference. In X-linked dominant inheritance, although the affected female transmits the trait to half her sons and half her daughters, the affected male transmits it to *none* of his sons, and *all* of his daughters are affected (they are not merely *carriers* as with recessive X-linked traits). One of the best studied X-linked dominant traits is vitamin D-resistant rickets, or hypophosphatemic rickets. The skeletal defect alone gives a pedigree pattern that tends to be inconclusive (Fig. 3.10a). However, when low blood phosphate is used as the trait for analysis the inheritance becomes clear (Fig. 3.10b). In all family studies the genetic analysis tends to be more precise the closer the phenotype studied is to the primary gene action.

On the average, the degree of severity of an X-linked dominant trait tends to be greater and to be less variable in affected males than in

FIG. 3.10. *Hypophosphatemic (or vitamin D-resistant) rickets. (a) When skeletal deformity is used as the phenotype, the pedigree pattern is unclear. (b) When low serum phosphate is used as the phenotype, the pedigree pattern is clearly that of an X-linked dominant trait. After T. F. Williams* et al., *in* The Metabolic Basis of Inherited Disease, *J. B. Stanbury, J. B. Wyngaarden, and D. S. Fredrickson, eds. New York: McGraw-Hill Book Co., 1960.*

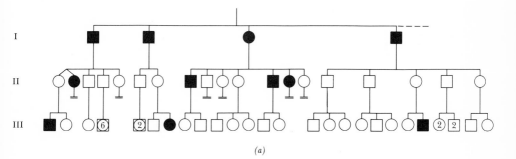

(a)

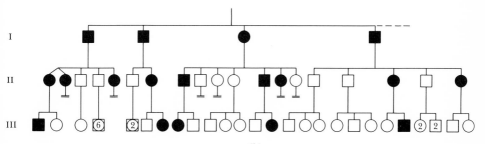

(b)

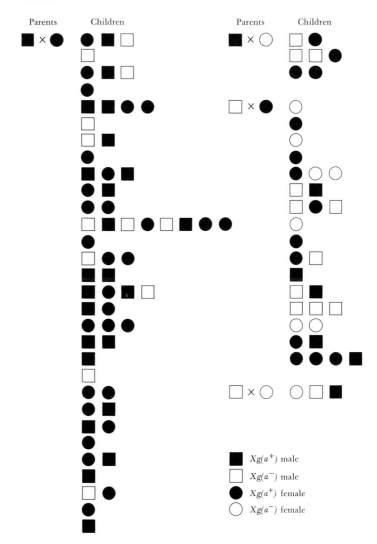

FIG. 3.11. *Family data on Xg blood group. From P. A. Moody (1967), based on data of J. D. Mann et al., Lancet, 1 (1962), 8.*

affected females. An explanation is provided by the Lyon hypothesis (p. 103). Note that the skipped generations in the pedigree of Fig. 3.9*a* involve in each case a female.

As is discussed on p. 127, rare X-linked dominant traits such as vitamin D-resistant rickets tend to be twice as frequent in females as in males. When a particular trait is frequent in a population, constituting a polymorphism, special methods must be used to establish mode of inheritance (p. 129). The distribution of a frequent X-linked dominant trait, the Xg^a blood group, in a randomly selected group of families, is shown in Fig. 3.11. It will be noted that more females than males are "affected," that is, are $Xg(a+)$. A critical mating is $Xg(a+)$ male by $Xg(a-)$ female. All sons are $Xg(a-)$; all daughters are $Xg(a+)$.

Effect of Sex on Gene Expression

Note the distinction between sex-linked (X-linked) inheritance and sex-influenced or sex-limited autosomal inheritance. Baldness appears to be a sex-influenced autosomal trait. In the male "pattern baldness" behaves as an autosomal dominant, but in the female baldness behaves as a recessive since the gene must be in homozygous state for baldness to occur in women. An exception to this is that baldness can occur in a heterozygous woman who develops a masculinizing tumor of the ovary. Baldness is a sex-influenced autosomal trait; the extreme case of sex influence is sex limitation.

It is difficult to distinguish X-linked recessive inheritance from sex-limited (that is, male-limited) autosomal dominant inheritance if the nature of the disease is such that reproduction by affected males cannot occur. Figure 3.12 shows the pedigree of a family in which many males are infertile because of a testicular disorder or more precisely a defect in response to testicular hormone, testosterone. Note that since the disorder can express itself only when testes are present (that is, only in the male), autosomal dominant and X-linked recessive (or dominant) inheritance account for the pedigree pattern equally well. An unequivocal example of sex limitation of an autosomal trait, in this instance

FIG. 3.12. *The pedigree of a family in which many males have a testicular disorder that renders them infertile. Is this an X-linked recessive or a male-limited autosomal dominant trait?*

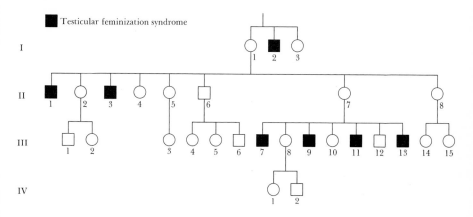

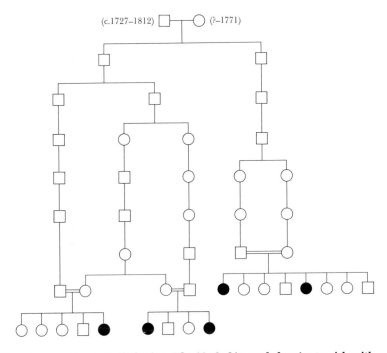

FIG. 3.13. *An example of sex limitation. The blacked-in symbols refer to girls with transverse vaginal septum. Autosomal recessive inheritance of this very rare condition is proved by the occurrence of affected girls in each of three sibships and by the fact that all six parents of affected girls share a common ancestral couple who lived in the 1700s. This autosomal malformation occurs only in females. From V. A. McKusick et al., J. Am. Med. Assoc., 204 (1968), 113–118.*

limitation to females, is provided by the condition in the family of which the pedigree is shown in Fig. 3.13. The disorder is transverse vaginal septum, an autosomal recessive malformation which has no counterpart in males.

The examples cited to this point illustrate Mendel's first law, that of segregation; traits are distributed in families as though the genes determining them segregate at meiosis. Alleles are alternative forms of genes that occur at the same genetic locus and determine alternative forms of the same trait. In several instances a considerable number of alternative genes have been identified at the same genetic locus, so-called multiple alleles. Any one individual can, of course, carry no more than two different alleles. In the ABO system, for example, the major alleles are A_1, A_2, B, and O (sometimes written I^{A_1}, I^{A_2}, I^B, I^O) ; these determine the particular blood group specificities that are designated by these letters.

Mendel's Law of Independent Assortment

Mendel's second law is used by the blood-group geneticist whenever he undertakes to prove that a newly discovered blood-group antigen represents a "new" system and is not merely part of a previously known one. Whereas segregation is the behavior of genes at the same locus (alleles), independent assortment is the behavior of genes at separate loci (nonalleles) (see Fig. 3.14). Alleles segregate; nonalleles assort.

Families of two or more children are needed to demonstrate independent assortment. The informative type of parental mating is the double backcross, in which one parent is heterozygous for both of the genes being investigated. The destination of this parent's genes in his children can be determined since the other parent has neither of them. The families diagrammed in Fig. 3.15 illustrate this approach. In these families two hemoglobin variants, Hb S and Hb C, occur. In the mating in which one parent has both and the other parent has neither aberrant hemoglobin, the first parent gives one or the other of the aberrant hemoglobins to each child. In no similar family observed to date has the "doubly affected" parent given both aberrant hemoglobins to a child and in none has he given neither to a child. Tentatively one can conclude from this evidence that the genes for Hb S and Hb C are allelic.

Figure 3.16 shows a different example from the hemoglobins. Again in this family, two variant hemoglobins occur, Hb S and a hemoglobin

FIG. 3.14. *The genetic consequences of allelism, linkage, and independent assortment.*

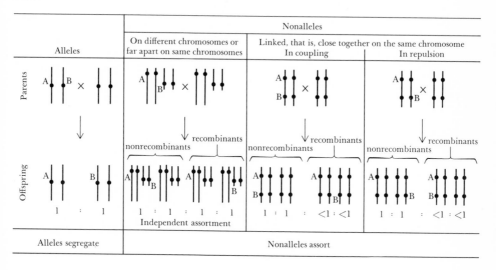

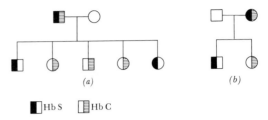

(a)   *(b)*

▮ Hb S   ▥ Hb C

FIG. 3.15. *Example of segregating alleles: the genes for hemoglobin S and hemo-globin C.* (a) *After E. F. Hays and R. L. Engle,* Ann. Internal Med., 43 *(1955), 412.* (b) *After E. W. Smith and J. R. Krevans,* Bull. Johns Hopkins Hosp., 104 *(1959), 17.*

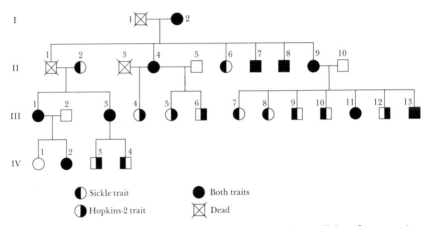

● Sickle trait   ● Both traits

◑ Hopkins-2 trait   ☒ Dead

FIG. 3.16. *An example of the independent assortment of nonalleles: the genes for hemoglobin S and hemoglobin Hopkins-2. Based on T. B. Bradley, S. H. Boyer, and F. H. Allen,* Bull. Johns Hopkins Hosp., 108 *(1961), 75.*

called Hopkins-2. In this instance, unlike the Hb S and Hb C example, a parent with both Hb S and Hb Ho-2 can give both or neither aberrant hemoglobin to an offspring (witness III 1 and II 9). Thus, these hemoglobins are clearly determined by nonallelic genes.

Chemical evidence can provide information on allelism and nonallelism. In light of the current one-cistron–one-polypeptide thesis (see Philip E. Hartman and Sigmund R. Suskind's *Gene Action* in this series) allelic genes would be expected to determine variation in the same polypeptide, whereas nonallelic genes would be expected to be concerned with different polypeptides. Such is indeed the case: both Hb S and Hb C, which by the pedigree data (for example, Fig. 3.15) are allelic, have a change in the β-polypeptide chain of hemoglobin; the difference of Hb Ho-2, which by the pedigree data (Fig. 3.16) is nonallelic with Hb S, resides in the α-polypeptide chain. The chemical and genetic analyses thus lead to the same conclusion.

For recessive genes also, nonallelism may be demonstrated by the familial pattern. Most genetic deaf-mutism (congenital deafness; deafness present from birth so that speech does not develop) is inherited as an autosomal recessive. Since phenotypically all such cases are identical, one might presume that genes at a single locus are responsible for all autosomal recessive deaf-mutism. Deaf-mute persons frequently marry deaf-mutes and the occurrence of pedigrees such as that in Fig. 3.17 demonstrates nonallelism. Specifically, since the children of the two deaf-mutes III 7 and III 9 had normal hearing, then the genes determining deaf-mutism in the part of the pedigree on the left (*a*) must be at a different locus than those determining deaf-mutism in the right-hand part (*b*). All the children of III 7 and III 9 are heterozygous at both loci, that is, "doubly heterozygous."

Another approach to the problem of allelism and nonallelism involves the frequencies of the several genotypes in the population. This method was used by Bernstein in 1925 to prove that the ABO blood groups are determined by a system of multiple alleles rather than by a pair of genes at one locus determining A and non-A blood type and a pair at another locus determining B and non-B blood type.

Linkage relationships of genes can give evidence of nonallelism. For example, two varieties of hemophilia (hemophilia A, or classic hemophilia, and hemophilia B, or Christmas disease) are, by the evidence of pedigree pattern, X-linked. One might wonder if the genes responsible for the two varieties are allelic. Such is clearly not the case because hemophilia A is closely linked with color blindness, whereas hemophilia B is determined by a gene located a considerable distance

---

FIG. 3.17. *Deaf-mutism in a family observed in Northern Ireland. The condition present in the three generations of (b) is apparently due to the same gene, but the deafness in (a) is due to a nonallelic gene, since marriage of presumed homozygotes from the two lines (a, b) resulted in no affected offspring. After A. C. Stevenson and E. A. Cheeseman, Ann. Human Genet., 20 (1956), 177–231.*

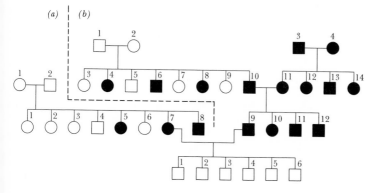

from the locus for color blindness. The linkage method can prove non-allelism, but because of its crudity cannot prove allelism in man (see p. 75).

Linkage

Linkage is the occurrence of two loci on one chromosome sufficiently close together so that something less than completely independent assortment takes place. If two loci are on separate nonhomologous chromosomes, then independent assortment occurs. Even if the two loci are on the same chromosome, if they are sufficiently far apart, crossing over can result in independent assortment. See Fig. 3.14. Traits determined by genes at loci rather close on the same chromosome tend to be transmitted together from generation to generation. The principles of linkage are also discussed in Franklin W. Stahl's *The Mechanics of Inheritance,* in this series.

The questions asked in linkage studies are the following: (1) Are the genetic loci occupied by the genes responsible for two traits on the same chromosome (or pair of chromosomes)? (2) If so, how far apart are the two loci? In practice the second question is answered first and the answer to the first question may follow directly. The distance between two loci determines the amount of crossing over that goes on between them. A measure of crossing over is provided by the proportion of recombinant or crossover individuals among the offspring of informative matings. If crossing over occurs to such an extent that 10 percent of offspring of informative matings are of the recombinant type, then the two loci are said to be about 10 map units apart. If crossing over occurs to the extent that 50 percent of offspring are of the recombinant type—a situation equivalent to independent assortment of Mendel's second law—then the two loci may be on different nonhomologous chromosomes or may be so far apart on the same long chromosome that independent assortment occurs through crossing over.

The analysis of linkage can be illustrated with the example of the Lutheran blood group (see p. 106) and secretor trait (which is the secretion of ABO blood-group substance into the saliva; see p. 107). Families are collected in which the parental mating is of the double backcross type (one parent is doubly heterozygous and one parent is homozygous recessive). Four types of offspring are possible: Lutheran positive and secretor; Lutheran negative and nonsecretor; Lutheran positive and nonsecretor; Lutheran negative and secretor. Note that even if the Lutheran and secretor loci are linked, no information is available at the start on whether the doubly heterozygous parent is in coupling or repulsion, that is, has the dominant gene on the same chromosome (coupling) or opposite chromosome (repulsion). The offspring are lined up as shown in Table 3.1. Although the four types of offspring are represented in about equal proportions in the series as

TABLE 3.1. *Data on Linkage of Lutheran and Secretor*

Parents:				
Phenotypes:	Lutheran positive, secretor		×	Lutheran negative, nonsecretor
Genotypes:	$Lu^aLu/Se\ se$			$Lu\ Lu/se\ se$

Offspring:				
Genotypes:	$Lu^aLu/Se\ se$	$Lu\ Lu/se\ se$	$Lu^aLu/se\ se$	$Lu\ Lu/Se\ se$
Phenotypes:	Lutheran positive, secretor	Lutheran negative, nonsecretor	Lutheran positive, nonsecretor	Lutheran negative, secretor

Sibship:				
1	0	0	6	2
2	4	1	0	0
3	0	0	4	1
4	1*	0	1	5
5	0	1	0	1*
6	1	1	0	0
7	1*	1*	1	1
8	0	0	3	3
9	4	1	0	0
10	0	3	0	1*
11	1	3	0	1*
12	2	2	0	0
13	0	0	1	1
14	1	2	0	0
15	1	2	0	1*
16	1	1	0	2*
TOTAL	17	18	16	19

SOURCES: *Sibship 1, J. Mohr (1951)*; *sibships 2, 3, J. Mohr (1953)*; *sibship 4–6, R. R. Race and R. Sanger (1958)*; *sibships 7, 8, S. D. Lawler and J. H. Renwick (1959)*; *sibship 9, M.N.Metaxas et al. (1959)*; *sibships 10–16, J.J.Greenwalt (1961)*.
* Probable crossovers. In sibships 5, 7, and 16 the crossovers are arbitrarily designated. Total number of offspring is 70. Total number of crossovers is 9. Recombination fraction is 13 per cent.

a whole, the distribution in individual families is far from equal. Those families with offspring only or predominantly to the left of the vertical line have the doubly heterozygous parent in coupling; the Lu^a and Se genes are on the same chromosome of a particular pair. Those families with offspring only or predominantly to the right of the vertical line have the doubly heterozygous parent in repulsion; the Lu^a and Se genes are on opposite chromosomes of a particular pair of homologues. Those individuals who fall on the side of the line opposite the majority are recombinants. In some cases, however, it is not certain which are

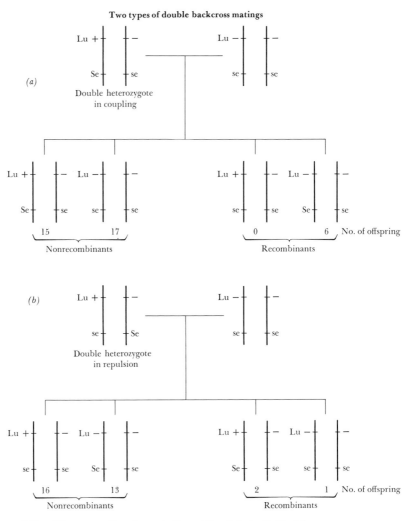

FIG. 3.18. *Diagrammatic representation of the data in Table 3.1.*

recombinants and which are nonrecombinants, for example, in families 5, 7, and 16. The data are presented in a different manner in Fig. 3.18.

Mapping of Loci on the X Chromosome

Analysis for autosomal linkage is difficult, both for the novice to understand and for the investigator to carry through in practice. X-linkage is somewhat more easily studied.

In studies of X-linkage the first of the two questions listed above (p. 66) is usually already answered in the affirmative—both loci are known from pedigree patterns to be on the same chromosome, the X. (This is not always the case, however, because male-limited autosomal dominant traits have a pedigree pattern identical to that of X-linked recessive traits; see p. 61.) The answer to the question of how far apart the loci are is provided by the phenotype of the sons of doubly heterozygous women. The one X chromosome of each son comes from the mother. X-chromosome crossing over can occur, of course, only in females. The father is irrelevant to the study of X-linkage—a circumstance fortunate for the study of populations with a high illegitimacy rate.

Male offspring of doubly heterozygous women will be of four types as indicated in Fig. 3.19, two noncrossover types and two recombinant or crossover types. If 2 out of 10 sons (20 percent of sons) of such women are of the recombinant type, then it is concluded that the two loci are separated by about 20 map units. But how does one tell the recombinants from the nonrecombinants? As should be clear from Fig. 3.19, one must know the coupling phase—whether the two traits are in coupling or repulsion, whether the responsible genes are on the same or opposite X chromosomes in the doubly heterozygous mother—before the recombinants and nonrecombinants can be so labeled.

The coupling phase in the mother is determined from the phenotype of *her* father, as illustrated in Fig. 3.19. Here, of course, illegitimacy

FIG. 3.19. *The grandfather method for mapping the X chromosome.*

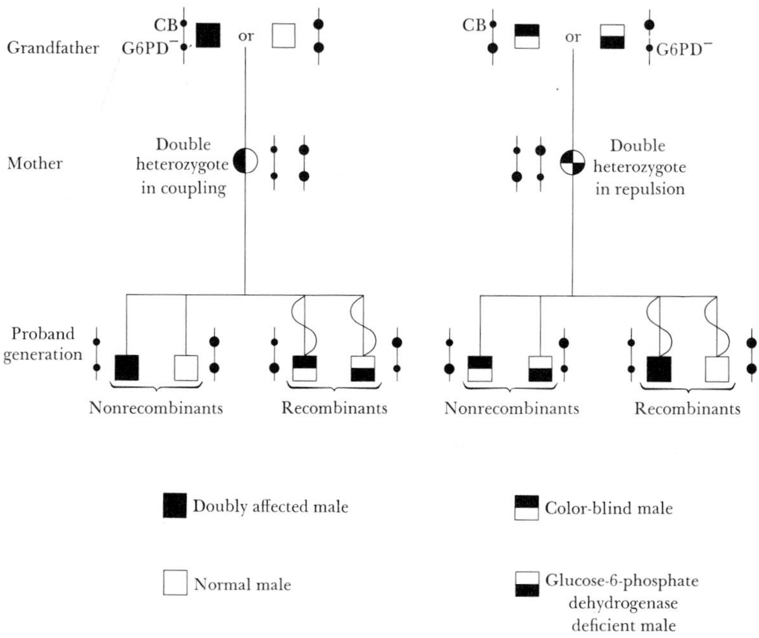

and the practical matter of availability of the mother's father for study do become important considerations.

The so-called Grandfather Method for measuring the distance between X-borne loci can be illustrated with an actual study of the linkage of color blindness and deficiency of glucose-6-phosphate dehydrogenase (G6PD) in the red blood cell. Negro schoolboys were first screened for color blindness. Then all males in the sibship of the color-blind boys were tested for G6PD deficiency. Thereby, sibships were ascertained in which both X-linked defects occurred. In the great majority of these the mother was doubly heterozygous. (In only a minority was the mother homozygous for one or both traits. These uninformative families can be excluded by testing the phenotype of the mothers with regard to these recessive traits.) Next the coupling phase of the mother was determined in each instance from the phenotype of the maternal grandfather of the proband. Determining the recombination fraction was then merely a matter of counting up the proportion of recombinant individuals among the sons of the doubly heterozygous women. The pedigrees shown in Fig. 3.20 are some of those actually found. Only one instance of recombination is shown. The loci for color blindness and for G6PD deficiency are rather close together on the X chromosome, probably about 5 map units apart.

The locus of the gene which determines classical hemophilia (also called hemophilia A) is closely linked with both the G6PD locus and the color vision locus. Other X-linked loci whose possible linkage with the color blindness and G6PD loci has been checked (these include another type of hemophilia called hemophilia B, or Christmas disease) have been found *not* linked. This may seem paradoxical. All are known to be on the X chromosome and in that sense are linked. However, some are so far apart that free recombination occurs through crossing

FIG. 3.20. *Some pedigrees in which the linkage between color blindness and glucose-6-phosphate dehydrogenase deficiency was studied. There is one definite crossover. Can you identify it? Based on I. H. Porter, J. Schulze, and V. A. McKusick, Ann. Human Genet., 26 (1962), 107.*

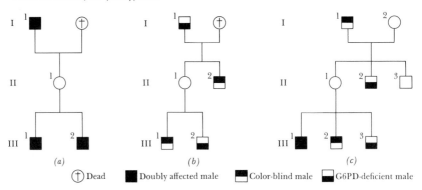

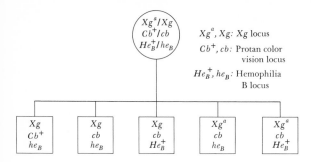

over. From other evidence the length of the X chromosome is such that nonlinkage of many loci is not surprising. Indeed, in one interesting family double crossing over seems to have occurred in at least one individual (see Fig. 3.21). The mother was heterozygous at the following 3 loci: (1) that for the X-linked blood group (Xg); (2) that for color blindness; and (3) that for hemophilia B. Ignoring relationships to the centromere, three possible sequences of the loci are possible— Xg-cb-He_B, Xg-He_B-cb and cb-Xg-He_B—and for each of these, eight possible coupling phases: Xg^a-$Cb+$-He_B^+, Xg^a-$Cb+$-he_B, Xg^a-cb-He_B^+, Xg^a-cb-he_B, Xg-$Cb+$-He_B^+, Xg-$Cb+$-he_B, Xg-cb-He_B^+, and Xg-cb-he_B (considering only one X chromosome and only one of the three possible sequences). Thus, there are 24 possible genotypes for the triply heterozygous mother. Whichever is correct, at least one of the sons must have resulted from a double crossover.

Difficulties in Detecting Autosomal Linkage

The problems in detecting and quantitating autosomal linkage in man account for the slowness with which mapping of the human chromosomes has proceeded. Some of these problems are: (1) The traits for study must be monogenic and uncomplicated in their genetics, with no ambiguity in the scoring of "affected" versus "unaffected." (2) The traits for study must be dominant. (3) At least one of the traits must be frequent so that doubly heterozygous parents are found fairly often. (4) Man has 22 pairs of autosomal chromosomes. If each of the available marker loci were on a separate chromosome, there would still not be enough "to go around." Actually some of the long chromosomes can have independent assortment of widely separated loci through crossing over. From chiasma counts performed on the meiotic chromosomes it is estimated that the total genetic length of the autosomes is about 2,700 map units (p. 36). (5) The coupling phase of doubly heterozygous persons often is not known, although sometimes it may be inferred from the rest of the pedigree. Scarcely more than a dozen traits fulfill the requirements to qualify as marker traits. Various methods by which polymorphic traits are demonstrated are presented on p. 89. The markers useful in linkage work include several separate blood-group systems (for example, ABO, Rh, Lutheran, MNSs, Duffy, Kidd, Kell), several serum protein polymorphisms (haptoglobin, transferrins, gamma globulin), and the ability to taste phenylthiocarbamide (PTC). The methods used in testing for marker traits applicable to linkage studies are reviewed in Chap. 4 (p. 89).

Because of the difficulties in detecting autosomal linkage in man, few such linkages are known (see Table 3.2). Note that the largest number are linkages between a rare abnormality and a common marker trait (No. 2, 3, 5, 7, 8, 10, and 11). Numbers 1 and 9 involve two marker traits, and No. 8 involves two rare abnormalities. The "confidence limits" on the estimate of the interval which separates the two linked loci are wide. For three of the most extensively studied linkages, approximate values for the interval are given.

Relationship of Recombination to Chiasmata

The relationship between genetic recombination and chiasmata is such that one chiasma is equivalent to a genetic length of 50 map units, arbitrarily defined. (By the time of crossing over, the synapsed chromosomes have replicated, that is, they are in the form of a tetrad. Two chromatids of the tetrad participate in crossing over and among the products of one crossover, half will be recombinants.) In the autosomes of the male about 54 chiasmata are counted. Thus, the total genetic length of the autosomes is estimated to be 54×50, or 2,700 map units. Since the X chromosome is about 6 percent the length of the haploid autosome set, the genetic length of the X chromosome is more than 150 map units. Thus, the genetic length of the X chromosome is sufficient that the finding of independent assortment of many pairs of loci tested to date (see p. 70) comes as no surprise.

The above estimate of the genetic length of the human X chromosome can be only an approximation because it is based on chiasma counts in the *autosomes* of the *male*, whereas the specifically relevant datum is the chiasma frequency in the X chromosomes of the female. In the mouse a somewhat higher frequency of recombination has been demonstrated in females than in males in the instance of several (although not all) linkages examined from this point of view. In man also a higher recombination frequency has been demonstrated in females (Table 3.2).

Note in Table 3.2 that one three-point linkage map has been es-

TABLE 3.2. *Autosomal Linkages*

---

1. Secretor factor locus and Lutheran blood group locus—about 13 map units (female 16; male 10).
2. ABO blood group locus and nail-patella syndrome locus—about 11 map units (female 14; male 8).
3. Rhesus blood group locus and one elliptocytosis locus—about 3 map units.
4. Beta and delta hemoglobin loci—contiguous.
5. Duffy blood group locus and a cataract locus—very close.
6. Transferrin locus and a pseudocholinesterase locus—about 16 map units (female 19; male 12).
7. Albumin (bisalbuminemia) locus and group-specific component (GC) locus—about 2 map units.
8. Pelger-Huet anomaly locus and a muscular dystrophy locus—very close.
9. ABO blood group locus and adenylate kinase locus—about 16 map units (female 24; male 8).
10. MNSs blood group locus and a rare scleroatrophic and keratotic disorder of the skin of the limbs—about 4 map units.
11. Adenylate kinase locus and nail-patella syndrome locus—very close.
12. Duffy blood group locus and a chromosome 1 "uncoiler" factor—about 3 map units.

---

tablished in man. The ABO locus, the nail-patella syndrome locus, and the adenylate kinase locus are linked and are probably situated in that sequence. Which chromosome carries these three linked loci is not yet known. However, one pair of linked loci—those for Duffy blood group (see p. 106) and a type of congenital cataract—appear to be on chromosome 1. This conclusion is based on the following two findings: (1) family studies showed that the Duffy blood group and a form of cataract are linked; (2) one of the chromosomes 1 in some normal individuals has an extended or uncoiled segment (see Chap. 2, p. 12, and Fig. 2.4). (It seems likely that an "uncoiler" factor carried on the chromosome which shows the extended segment is responsible.) When conventional studies for linkage between chromosome 1 and various marker loci were performed in families in which the marker chromosome was segregating, the Duffy blood group locus was found to be linked to the anomalous chromosome 1, or more precisely, perhaps, to the "uncoiler" factor. Thus the Duffy locus and the locus for a form of cataract are presumably located on chromosome 1.

Other Approaches to the Demonstration of Linkage

The close linkage of the beta and delta loci, which determine the beta and delta chains of hemoglobin (see p. 95), is indicated not only by classic family studies but also by three other types of evidence: (1) The beta and delta chains have close chemical similiarities (in fact only 10 amino acids out of 146 are different), suggesting that they arose from a common ancestor through gene duplication and that they are probably contiguous. (2) A bizarre aberrant hemoglobin has been found, hemoglobin Lepore, in which the nonalpha polypeptide in part resembles the beta chain and in part the delta chain. Presumably the new gene (for Hb Lepore) arose through unequal crossing over that involved the two contiguous loci (see p. 40 and Fig. 2.20*b*.) (3) Several rare mutant alleles of the delta gene have been found. In all cases in which a delta mutant allele has occurred in the same individual as the sickle gene, a beta-gene mutant, the two genes have been in repulsion. This is interpreted as indicating very close linkage and insufficient time in generations for equilibration of the coupling and repulsion phases.

The methods just outlined can be resorted to in human genetics as substitutes for family data that are difficult to assemble in a volume that provides a critical answer on the question of linkage. Comparative studies also have usefulness. In mice the loci which determine the gamma and beta chains of hemoglobin are closely linked. Although no direct evidence is available, the demonstrated tendency for close linkages to persist during the course of evolution would suggest that these loci are linked in man also.

Association

Genetic linkage is quite different from blood-group and disease association. Association is the nonrandom occurrence of two genetically separate traits in a population. Before enough generations have passed for a chromosome to be minced up by the process of crossing over, association on the basis of genetic linkage may be observed, but linkage produces no permanent association in the population and most

TABLE 3.3. *Lutheran and Secretor Phenotypes in 400 Unrelated Persons*

Phenotype	Lu (a+)	Lu (a—)
Secretor	27	270
Nonsecretor	8	95
TOTAL	35	365

SOURCE: S. D. Lawler and J. H. Renwick, "Blood Groups and Genetic Linkage," *Brit. Med. Bull., 15* (1959), 145–49.

TABLE 3.4.   *Association of Duodenal Ulcer and Blood Type O**

Phenotype	Ulcer	Nonulcer
Type O	535	4,578
Type A	311	4,219
TOTAL	846	8,797

SOURCE: J. A. Fraser Roberts, *Brit. Med. Bull.*, 15 (1959) , 129.
* Relative risk of type O persons:

$$\frac{535 \times 4{,}219}{311 \times 4{,}578} = 1.59 \qquad \chi^2 = 0.16$$

A similar calculation from the secretor-Lutheran data indicates no significant association:

$$\frac{27 \times 95}{8 \times 270} = 1.19 \qquad \chi^2 = 37$$

association has its basis in mechanisms other than genetic linkage. For example, the Lutheran blood-group locus and the secretor locus are known to be rather closely linked, being separated by about 13 cross-over units (Table 3.1). Yet in any group of unrelated persons one Lutheran blood type does not occur more frequently with secretor than with nonsecretor. See Table 3.3, which shows that about 10 percent of both secretors and nonsecretors were Lutheran positive. Furthermore, in the linkage data presented in Table 3.1, the numbers of persons with both traits, one trait only, or neither trait are about equal. No linkage is indicated and linkage becomes evident only when the individual families are studied. Conversely, blood group O and peptic ulcer of the duodenum show significant association (see p. 190). This is due to some physiologic peculiarity of the type-O person that predisposes him to peptic ulcer and is not due to genetic linkage. Table 3.4 gives data on the association of blood group O with duodenal ulcer. This may usefully be contrasted with Table 3.3. The calculations accompanying the table illustrate the cross-multiplication method of evaluating relative risk.

As is further discussed on p. 88, genetic linkage is also quite different from syndromal relationship.

Close Linkage versus Allelism

In human pedigrees close linkage is difficult to distinguish from allelism. The number of progeny from informative matings is small compared to the numbers on which recombination estimates can be based in experimental species. The problem arises in connection with the variant hemoglobins and the Rh and some other blood groups (p. 107).

A Cytologic Demonstration in Man of the Laws of Inheritance

As a synthesis and summary of this chapter and Chap. 2, it is useful to review the findings in connection with a "marker" chromosome. The cytologic basis of Mendel's laws of segregation and independent assortment are demonstrated and, potentially, linkage can establish one or more loci situated on the marker chromosome.

In the Old Order Amish community of Lancaster County, Pennsylvania, about 10 percent of persons have been found to carry a giant-satellited Group D chromosome (Fig. 3.22), which by autoradiographic labelling pattern appears to be chromosome 14.

Mendel's first law, that of segregation of alleles, can be illustrated as shown in Fig. 3.23a and 3.23b. The somatic chromosomes, schematized at the top in their familiar appearance at metaphase, undergo two stages of meiosis in the production of the egg and sperm. The two chromosomes of a given pair segregate, that is, go into separate daughter cells, in the first or reductional stage. The individual heterozygous for the marker chromosome produces haploid gametes of two types: those with the marker chromosome 14 and those with the usual chromosome 14. When married to a "normal" person, the heterozygous person will produce offspring of two types (normal and heterozygous) in equal proportions (Fig. 3.23a). Among 82 offspring of heterozygote × normal matings tested to date, 47 are heterozygous. (A deviation from the 50 percent predicted from theory by as much as this or more has a probability of 20 percent. Since many of the matings were taken into consideration because of the finding of the giant-satellited chromosome in an offspring, the proband may be excluded from the calculation, with resulting closer fit with the expected 50 percent.)

Segregation is also illustrated by the results of the mating of heterozygote with heterozygote (Fig. 3.23b). Each parent produces two classes of gametes and these can combine into three types of offspring: homozygotes for marker chromosome 14, heterozygotes for marker chromosome 14, and normal for chromosome 14, in the proportions of 1 : 2 : 1. This mating, which corresponds to the F_1 intercross of experimental genetics, been observed once in the Amish (Fig. 3.23c). The offspring number eight, among whom the exact Mendelian ratio is realized! (Among eight-sib families from a mating of heterozygote × heterozygote the exact 1 : 2 : 1 ratio is expected in slightly over 10 percent.)

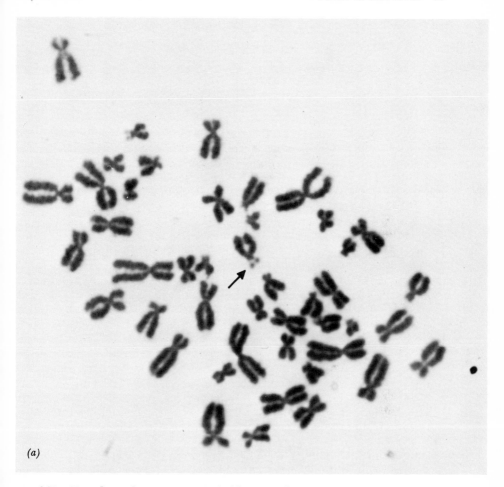

(a)

FIG. 3.22. *Metaphase chromosomes of Amish male showing one No. 14 with giant satellites (arrow). (a) Metaphase plate. (b) Group D chromosomes.*

(b)

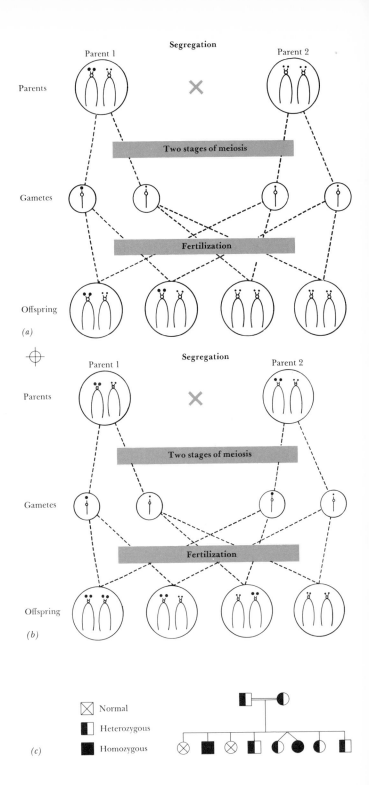

Segregation

Parent 1

Parent 2

Parents

Two stages of meiosis

Gametes

Fertilization

Offspring

(a)

Segregation

Parent 1

Parent 2

Parents

Two stages of meiosis

Gametes

Fertilization

Offspring

(b)

☒ Normal

▮ Heterozygous

■ Homozygous

(c)

FIG. 3.23. (a) *Segregation (schematic) in a mating of heterozygote by homozygote.* (b) *Segregation (schematic) in a mating of heterozygote by heterozygote (F_1 intercross).* (c) *Actually observed intercross.*

Mendel's second law, that of independent assortment of nonalleles, is illustrated by the Amish marker chromosome when one focuses on the transmission of the marker chromosome in relation to the transmission of the Y chromosome. The informative mating is that of a male with the marker chromosome ("double heterozygote," if you will) times a normal female. As schematized in Fig. 3.24 the male produces four types of sperm. Therefore on fertilization of the normal egg, four classes of offspring result. Among the Amish, 41 offspring from 7 informative matings have been tested. The proportions are as shown in Table 3.5. This much deviation from the expected $1:1:1:1$ proportion would result in somewhat more than 6 percent of samples of this size. Again, bias of ascertainment (See Chapter 6) may account for the excess of offspring carrying the satellited chromosome.

Family studies for genetic linkage between the giant satellite and loci for particular blood groups or serum protein types are under way. Demonstration of linkage would establish that these loci are situated on chromosome 14. Figure 3.25 schematizes the transmission of genes on the long arm of chromosome 14 according to whether they are close to the centromere (in *a*) or far from it (in *b*).

Let us use the ABO blood group system by way of example. The particularly informative mating in linkage is that of doubly heterozygous parent by doubly homozygous parent—in this instance, for example, a mating in which one parent is heterozygous for the marker chromosome and also blood group AB and the other parent is homozygous for the normal chromosome 14 and for blood group O.

During the first stage of meiosis, when the two chromosomes of a given pair have come into juxtaposition through the process of synap-

TABLE 3.5 *Offspring of Heterozygous Male by Homozygous "Unaffected" Female*

		(Male) Y	(Female) No Y
Giant satellited group D chromosome	Yes	17	11
	No	6	7

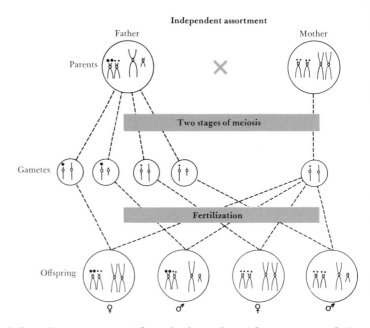

FIG. 3.24. *Independent assortment (schematic) in mating of heterozygous male by homozygous female.*

sis and when each chromosome is in a duplicated, two-strand form so that the combination is called a tetrad, crossing over occurs between strands of homologous chromosomes. The site of crossing over occurs at random in the length of the long arm, but the farther apart two loci are situated on the long arm, the greater is the chance that crossing over will occur between them. If the ABO locus is, for example, at the end of the long arm distal from the centromere (see Fig. 3.25b), then crossing over will occur with maximal probability. Four types of gametes are produced by the doubly heterozygous parent. Two of them (shown on the left and right sides) are nonrecombinants—the giant satellite and blood group A are on one chromosome and the normal satellite and group B are on the other chromosome, as in the doubly heterozygous parent. Two classes of gametes are recombinant— through crossing over the giant satellite and blood group B have ended up on the same chromosome and the normal satellite and blood group A on the same chromosome. Because the crossing over occurs

FIG. 3.25. (a) *Tight linkage (schematic): double backcross mating. Note that few recombinants would be observed.* (b) *Loose linkage (schematic): double backcross mating. Note that independent assortment might be observed, that is, half the offspring might be of the recombinant type.*

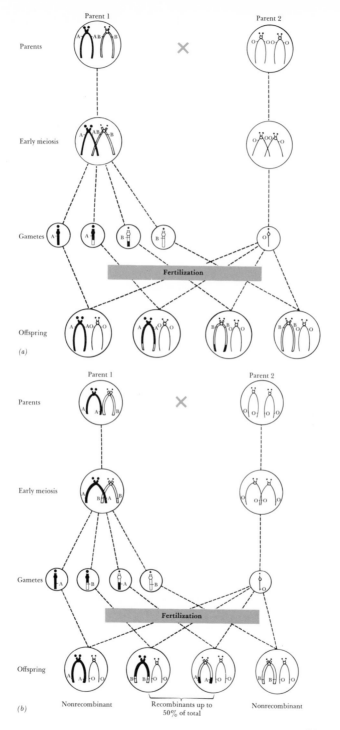

at a four-strand stage, the maximum recombination fraction is 50 percent—half the gametes can be of the recombinant type. This is equivalent to independent assortment of genes at loci on separate chromosomes (Fig. 3.24). Linkage studies of practicable size cannot distinguish between loci on separate chromosomes and loci far apart on a single chromosome. Note that the fact that the other parent is doubly homozygous insures that the phenotype of the offspring indicates whether recombination has occurred or not.

Note also that in some other doubly heterozygous persons the blood group B gene might be on the same chromosome as the giant satellite so that in his offspring B would tend to segregate with the giant satellite. In the general sense linkage is a phenomenon of *loci,* not of specific *alleles* at loci. The blood group A gene and the giant satellite are said to be in *coupling* if on the same chromosome, and in *repulsion* if on opposite chromosomes of the homologous pair.

If the ABO locus is at intermediate locations on the long arm of chromosome 14 between the centromere and the free end, the amount of recombination with reference to the giant satellite will measure its

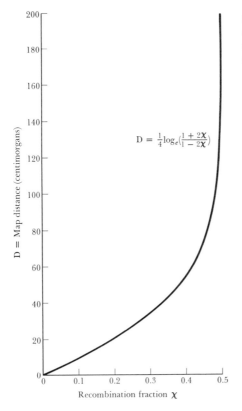

$$D = \frac{1}{4}\log_e\left(\frac{1 + 2\chi}{1 - 2\chi}\right)$$

D = Map distance (centimorgans)

Recombination fraction χ

FIG. 3.26. *The relationship between recombination fraction (χ) and physical map distance in centimorgans. After Kosambi; courtesy of E. A. Murphy.*

relative position. The long arm may be sufficiently long that as many as two chiasmata (the cytologic counterparts of genetic crossing over) may on the average form on it, in which case the genetic length of the long arm is said to be 100 map units, recombination units, or centimorgans. It follows that the relation between recombination fraction and distance separating two loci on a chromosome is not linear. The recombination fraction can go up only to 50 percent whereas the genetic length (and physical length) may be considerably more. The nonlinear relationship between recombination fraction and physical length is shown in Fig. 3.26. Note that the recombination fraction does not increase as fast as the physical length (or true genetic length) because, as the interval between two linked loci increases, the chance that not one but two crossovers will occur between them increases, and two crossovers is the equivalent of no recombination.

How many crossovers can one have in the long arm of chromosome 14? Since the genetic length of the autosomes is about 2,700 map units (p. 72) and since chromosome 14 represents about 3.9 percent of the total length of the haploid set of autosomes, one concludes that its genetic length is about 100 map units. Thus, there is ample opportunity for loci in the distal part of the long arm to assort independently of the giant satellite.

References

Donahue, R. P., W. B. Bias, J. H. Renwick, and V. A. McKusick, "Probable Assignment of the Duffy Blood Group Locus to Chromosome 1 in Man," *Proc. Natl. Acad. Sci. U.S., 61* (1968), 949–55.

Lawler, S. D., and J. H. Renwick, "Blood Groups and Genetic Linkage," *Brit. Med. Bull., 15* (1959), 145–49.

McKusick, Victor A., "On the X Chromosome of Man," *Quart. Rev. Biol., 37* (1962), 69–175; also, American Institute of Biological Sciences monograph, 1964.

Moody, P. A., *Genetics of Man.* New York: W. W. Norton & Co., Inc., 1967.

Renwick, J. H., "Elucidation of Gene Order," in *Recent Advances in Human Genetics,* Lionel S. Penrose and H. L. Brown, eds. Boston: Little, Brown & Co., 1961, pp. 120–38.

Renwick, J. H., "Progress in Mapping Human Autosomes," *Brit. Med. Bull., 25* (1969), 65–73.

Genes in the Individual

Physiologic genetics is concerned with the way genes work, in collaboration with extrinsic, or environmental factors, to determine the phenotype. Biochemical genetics is essentially the same as physiologic genetics, because the objective of physiologic genetics is to understand gene action in biochemical terms.

The genotype is the genetic constitution of the individual. The phenotype is the character or trait or the composite of characters that is capable of being observed but that may be many steps removed from the genotype. The distinction might be compared to that between character and reputation. Genotype and character are what one really is; phenotype and reputation are what one appears to be.

Because it is so far removed from the genotype, the phenotype is not necessarily an indication of the genotype. Environmental factors can result in the same phenotypic change as a mutant gene. *Phenocopy* was Richard B. Goldschmidt's term for such environmentally induced mimics. Furthermore, different genes can result in the same phenotype. *Genetic mimic* (or *genocopy*) is the term for this phenomenon. For example, some hereditary disorders display autosomal dominant inheritance in some families, autosomal recessive inheritance in others, and X-linked inheritance in yet others. Elliptocytosis (oval-shaped erythrocytes), an autosomal dominant trait, is determined in some families by a gene rather closely linked to the Rh blood-group locus, whereas in other families the seemingly identical phenotype is determined by a locus not linked to Rh.

Genetic heterogeneity is the term applied to the situation in which more than one genetic cause leads to the same or very similar phenotypes. In medical genetics in particular, it has been found repeatedly that what was thought to be a single entity has proved on close study to consist of two or more distinct entities. Sometimes the heterogeneity is allelic, sometimes nonallelic; that is, the distinct genes resulting in a similar phenotype may be at the same locus or at different loci.

A trait may be highly variable from one person to another. Autosomal dominant traits in particular tend to show pronounced variability. *Expressivity* is the term that has been applied to this characteristic of a hereditary trait, and is equivalent to "grades of severity" in clinical medicine. A population of persons of a particular genotype will fall into a bell-shaped normal distribution curve according to the expressivity of the hereditary trait determined by that genotype. A majority of "affected" persons have an intermediate grade of severity, whereas a few are severely affected and a few have mild involvement, perhaps sometimes so mild that the presence of the gene escapes detection. The curve is skewed in one direction or the other in the case of most traits. The basis for such variability in expression is partly environmental and partly genetic. The individual mutant gene primarily responsible for a trait does not operate in a vacuum, but rather against the background of the rest of the genome. In the case of autosomal dominant traits the presence of different wild-type alleles can modify the expression of the mutant gene. In the nail-patella syndrome, an autosomal dominant disorder, the grade of severity shows a much higher correlation between affected sibs than between offspring and affected parent. *Isoalleles* is the term applied to multiple wild-type alleles that appear to have identical phenotypic effects except when observed in heterozygous combination with a mutant allele, in this case that for the nail-patella syndrome.

Genetic traits that are subject to considerable modification by the effects of genes other than the one primarily responsible for the trait and also by environmental influences may not be recognizable in some individuals despite the fact that they have the gene (or gene pair) that in a majority of instances "causes" the trait. In such cases the trait (or the gene) is said to be nonpenetrant. Penetrance is an all-or-none proposition. Nonpenetrance occurs at the mild end of the curve of expressivity. One may think of the manifestation curve of the mutant gene as overlapping at the mild end of the expressivity curve in normals, or at least in those who do not carry the gene that is being studied (Fig. 4.1). In the area of overlap the persons with the gene cannot be distinguished from those without the gene. The threshold of penetrance, represented by a vertical line cutting the expressivity curve,

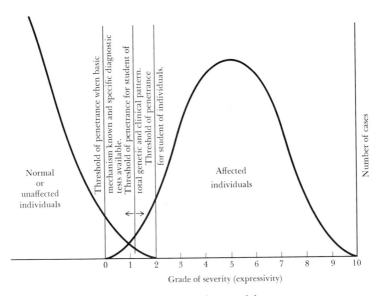

Normal or unaffected individuals

Threshold of penetrance when basic mechanism known and specific diagnostic tests available.
Threshold of penetrance for student of total genetic and clinical pattern.
Threshold of penetrance for student of individuals.

Affected individuals

Number of cases

Grade of severity (expressivity)

FIG. 4.1. *The concepts of penetrance and expressivity.*

moves progressively to the left as methods for analyzing the phenotype become more discerning. Penetrance and nonpenetrance often are functions of the acuteness of our methods of study. Figure 4.2 gives an example of the overlap of phenotypes when individual effects of a single gene pair, in this case that for phenylketonuria, are singled out for analysis. If hair color, head size, and intelligence were the only phenotypic characteristics available for distinguishing phenylketonuric patients from normals, the error in diagnosis would be considerable. The genotype could be said to be nonpenetrant in a great number of persons. But when the level of phenylalanine in the blood is used as the phenotype for study, the genotype is fully penetrant.

In medicine many hereditary disorders are syndromes. Meaning literally "a running together," this word refers to combinations of manifestations that occur together with reasonable consistency. Most genetic syndromes have their basis in a single mutant gene. The manifold features are often the consequence of the fact that the mutant gene and its wild-type counterpart have a widespread role in the body's economy. Or the effects of the mutant gene may be such that a substance toxic to several different tissues accumulates in the body. There are other mechanisms for multiple and often seemingly distinct manifestations of a single gene. At the phenotypic level, it is permissible to refer to the gene as *pleiotropic* (pronounced ply-o-tropic) in its action, that is, having multiple effects. *Polyphenic* is also a good term for describing the gene responsible for such syndromes, but one

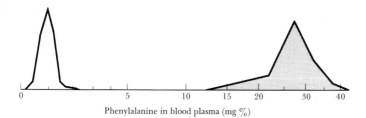

Phenylalanine in blood plasma (mg %)

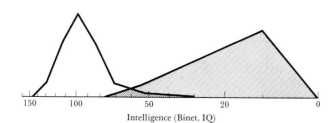

Intelligence (Binet, IQ)

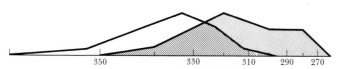

Head size. Length + breadth in mm (corrected for sex)

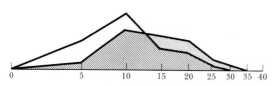

Hair color. Reflectance % at 700 mμ (corrected for age)

FIG. 4.2. *Frequency distributions of some characteristics of phenylketonuria (PKU) in PKU patients (shaded) and in control populations. Hair color and head size show pronounced overlap, and intelligence shows some overlap. The level of phenylalanine in the blood, however, is higher in all PKU patients than in controls. If intelligence were the only phenotype used in the analysis, the genotype would be said to be nonpenetrant in a small proportion of cases. When plasma level of phenylalanine is the phenotype, the genotype is found to be fully penetrant. Redrawn from L. S. Penrose, Ann. Eugenics, 16 (1951), 134.*

must remember that at the level of primary gene action there is no evidence that one gene has more than one function.

To a considerable extent, the individual components of a syndrome can vary independently. In some persons, for example, one component of the syndrome may be completely lacking. Experiments of L. C. Dunn and of others, involving observations on the effects of mutant genes on different genetic backgrounds in experimental species, give support to the above statement. In man numerous observations are best explained by genetic modification of individual components. For example, in the Marfan syndrome, in the same family some affected persons may have the full-blown syndrome of eye, skeletal, and aortic abnormalities, whereas others escape the eye abnormalities.

So far as is known, in no heritable syndrome of man is the association of traits the result of close linkage of several genes, each responsible for one of the aspects of the syndrome. That several linked genes would mutate simultaneously to reproduce a given syndrome with exactitude is unlikely. If all instances of the syndrome were the result of a unique event in the remote past, then the several genes would be likely to have become separated through the process of crossing over and no particular association of the several manifestations would be observable.

It is true that a genetic syndrome like Down's syndrome has its basis in other than a single gene. Presumably dosage effect of many genes carried on chromosome 21 and present in excess is responsible for the characteristic clinical picture of trisomy 21 (see p. 21). It is possible, furthermore, that certain very rare syndromes that are transmitted in a Mendelian manner are the result of small chromosomal aberrations, such as deletion or inversion, affecting the action of several genes.

Genetics: the Science of Variation

Environmental factors in variation are of secondary interest to the geneticist but can never be disregarded. Discontinuous variation, such that persons can be classified as having a given trait or not having the trait, is more easily studied, especially in man, than is continuous variation. Consequently, the amount of precise genetic information is greater for discontinuous traits than it is for continuous traits.

E. B. Ford of Oxford University has described polymorphic traits in a famous definition that is quoted like a verse of scripture: *"Polymorphism* may be defined as the occurrence together in the same habitat of two or more discontinuous forms of a species in such proportions that the rarest of them cannot be maintained merely by recurrent mutation."* The definition is so worded that geographic differences and those due to rare disease alleles constantly eliminated by natural selection and replaced by mutation are not included in the category of polymorphisms.

In man polymorphism has been demonstrated by a variety of methods for studying the phenotype:

(1) Physiologic methods. Color blindness satisfies Ford's definition. So does PTC-tasting—the ability or lack of ability to taste phenylthiocarbamide—although no complete separation between taster and nontaster groups is achieved.

(2) Morphologic methods. One has difficulty citing a morphologic trait that is relatively frequent and at the same time is a clearly "discontinuous form." Red hair and blue eyes tend to segregate as autosomal recessive characters. However, hair and eye color, attached or unattached ear lobes, clockwise or counterclockwise "cowlick" (hair whorl), and so on, have some features of continuous characters and are not readily amenable to simple Mendelian interpretation.

(3) Electrophoretic methods. Various types of electrophoresis—for example, on paper and in starch gel—have been used to demonstrate polymorphism of hemoglobin and of serum proteins (for example, the haptoglobins and transferrins). See Fig. 4.14 (p. 105) for a demonstration of haptoglobin types.

(4) Immunologic methods. The erythrocyte groups ("blood groups"), starting with the ABO system discovered by Landsteiner in 1900, represent the outstanding examples of immunologically demonstrated polymorphism. The gamma globulin groups (Gm) of human serum are also demonstrated by immunologic techniques.

(5) Immunoelectrophoretic methods, a combination of (3) and (4), have been used to demonstrate polymorphism of the α_2 globulins of serum (the Gc types) and of beta lipoproteins (the Ag types).

(6) Metabolic methods, which might be included in the category of physiologic methods, have been used to demonstrate polymorphism with regard to the rate of acetylation of isoniazid, a drug used in the treatment of tuberculosis.

(7) Enzymatic methods demonstrate differences in erythrocyte glucose-6-phosphate dehydrogenase (G6PD). Deficiency of this enzyme, an X-linked trait, occurs as a polymorphism in African and Mediterranean populations.

(8) Methods which might be called *enzymo-electrophoretic* are also used to demonstrate polymorphism. Examples include certain enzymes of the erythrocyte, such as acid phosphatase, phosphoglucomutase, and adenylate kinase.

New techniques for demonstrating discontinuous variation in man are much to be desired. Simple observational methods have limited usefulness. Few genetic markers for linkage studies—discontinuous traits with impeccable Mendelian transmission—are available. All science is enslaved to its methods. By his invention of starch gel electrophoresis, Oliver Smithies contributed greatly to human genetics as well as to protein chemistry. A number of genetic polymorphisms, for example, that of the haptoglobins, have been demonstrated with this technique.

Protein Structure and Gene Action

Earlier it was stated that one gene has one function. There is now good evidence that for many genes (so-called structural genes) the function is to specify the amino acid sequence of a protein or of one polypeptide chain. The protein so specified may be an enzyme; it may be a protein, such as hemoglobin or haptoglobin, with a special function; or it may be a structural protein such as collagen. Much of the present knowledge of gene action, especially the role of the gene in protein synthesis, is based on study of the sickle-cell condition, a polymorphism of man. Because this important area is discussed in full in another volume of this series, Hartman and Suskind's *Gene Action,* only a résumé will be provided here.

The history of the development of knowledge of sickle hemoglobin illustrates the various levels of sophistication in the analysis of phenotype. First, sickling (formation of a peculiar shape of the erythrocytes) and sickle-cell anemia were identified (Fig. 4.3a). From family studies, it was concluded that those persons in whom only sickling occurred were heterozygous, whereas persons in whom severe anemia accompanied sickling were homozygous. Then, Linus Pauling and his colleagues discovered that the hemoglobin in sickle-cell anemia patients has an electrophoretic mobility different from the hemoglobin of normal persons (Fig. 4.3b) and that the hemoglobin of heterozygotes is partly of the normal type (Hb A) and partly of the sickle type (sickle hemoglobin, or Hb S).

Thereafter it was found that the normal hemoglobin molecule is a tetramer, that is, is made up of four polypeptide chains: two identical alpha chains and two identical beta chains (Fig. 4.4). The peculiarity of sickle hemoglobin is a feature of the beta chains, with the alpha chains of Hb S being identical to those of Hb A.

Next it was demonstrated that the defect in sickle hemoglobin is limited to one peptide of the β-polypeptide chain (Fig. 4.4). Vernon Ingram applied to hemoglobin the technique of "fingerprinting." In this technique the polypeptide chains are broken up into peptide fragments by means of the enzyme trypsin. The mixture of peptides is then spread out on paper by electrophoresis followed by chromatography at right angles to the direction of the electrophoresis. The resulting display of the peptides is the fingerprint (Fig. 4.3c). The results of this method show that sickle hemoglobin differs from the normal with respect to one peptide. Peptide 4 is altered in its electrophoretic mobility.

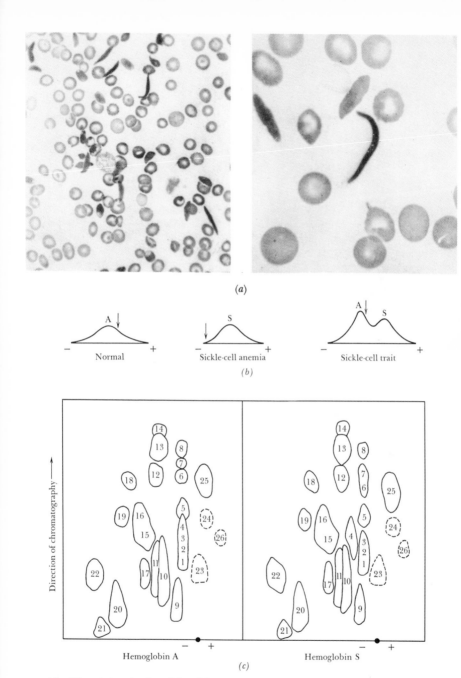

FIG. 4.3 *Phenotypes produced by sickle hemoglobin (Hb S). (a) A comparison of "sickled" erythrocytes with normal ones. (b) Free electrophoresis of the hemoglobin of a normal subject, homozygous AA; a person with sickle-cell anemia, homozygous SS; and one with sickle-cell trait, heterozygous SA. (c) "Fingerprint" of Hb A and Hb S.*

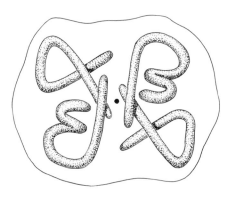

FIG. 4.4. *A schematic representation of the two alpha and two beta chains of the hemoglobin molecule. After V. M. Ingram,* Nature, *184 (1959), 1905.*

When the amino acid sequence of peptide 4 was determined in Hb A and Hb S, the following was found:

Hb A   valine-histidine-leucine-threonine-proline-
                 glutamic acid-glutamic acid-lysine-
Hb S   valine-histidine-leucine-threonine-proline-
                 valine-glutamic acid-lysine-

It is now known that the segment of the beta chain shown above is at the free amino ($-NH_2$) end and that the beta chain contains 146 amino acids, only one of which in sickle hemoglobin is different from the normal. The reason Hb S is slower in its electrophoretic mobility is that it lacks the negative charge of the glutamic acid residue in Hb A; valine, unlike glutamic acid, has no free charge.

Several other variant hemoglobins have been found to have an amino acid change in the beta chain. In one of these, Hb C, the very same amino acid is involved: lysine is substituted for glutamic acid as the sixth amino acid in the beta chain. All of the variant hemoglobins with changes in the beta chain appear to be determined by genes allelic with each other (see p. 63) .

Other variant hemoglobins, for example, Hb Hopkins-2, so-called because it was a second variant discovered at the Johns Hopkins Hospital, show a change in the alpha chain. On the basis of studies of families in which both an alpha-chain mutation and a beta-chain mutation are segregating, it has been concluded that the mutations are nonallelic; that is, that separate genetic loci determine the alpha and beta chains. The loci may even be on separate chromosomes.

The concept currently accepted is that genes act by determining the amino acid sequence of proteins, or rather, as the hemoglobin evidence shows us, polypeptide chains. This formulation is a logical outgrowth of Beadle and Tatum's one-gene–one-enzyme hypothesis of 1940. Clearly their hypothesis was too restrictive since the protein specified is not necessarily an enzyme; "one-gene–one-polypeptide" is a better statement of what is currently thought to be the case. The genetic locus is thought of as the unit that specifies the amino acid sequence of one polypeptide. The cistron, or functional unit, of Seymour Benzer can be considered synonymous; that is, it has the same physical limits as the locus. Codons are the subunits of the locus, each encoding the information for one amino acid. Presumably the mutations in Hb S and Hb C are at the same site, that is in the same codon. The mutations responsible for Hb E and other beta-chain variant hemoglobins are at the same locus but not at the same site. Recombination can occur within the locus so that the recon, or recombination unit, of Benzer is something less than the locus. It is theoretically possible, for example, that a child with only Hb A or a child with a hemoglobin containing two amino acid substitutions might occur among the children of an individual with both Hb S and Hb E, married to a person with only Hb A. Since the probability of crossing over, which could produce such a result, is directly related to the distance separating the reference points, crossing over between two codons as close as those involved in Hb S and Hb E would be very rare. Intracistronic recombination probably explains those aberrant hemoglobins which have *two* amino acid changes in the same polypeptide chain. Hemoglobin C, as mentioned earlier, has a substitution of lysine for glutamic acid as the sixth amino acid in the beta chain; hemoglobin Korle-Bu has a substitution of asparagine for aspartic acid as the seventy-third amino acid in the beta chain. Hemoglobin C (Harlem) contains *both* amino acid substitutions and may have arisen by intracistronic crossing over in a compound, that, is a person with the Hb C gene on one homologous chromosome and the Korle-Bu hemoglobin gene on the other. (Double mutation, for example, mutation in a Hb S gene, is an alternative possibility.)

The muton (mutational unit) and recon (recombination unit), in the terminology of Benzer, are even smaller than the codon. Assuming a triplet code for each amino acid, one can see that a change in only one of the three bases would represent a mutation. Furthermore, recombination can occur between two adjacent bases.

Information from studies of the genetic code in cell-free systems correlates well with information on the change in certain aberrant hemoglobins. For example: in RNA, either GAA or GAG codes for glutamic acid, the amino acid which is sixth from the $-NH_2$ end of the beta chain of Hb A. The change from Hb A to either Hb S or Hb C can occur through the change of only one base, inasmuch as the triplets GUA and GUG code for valine (present at position 6 in Hb S) and the triplets AAA and AAG code for lysine (present at position 6 in Hb C). See Table 4.1. (U = uracil, A = adenine, G = guanine, C = cytosine. In the DNA code thymine replaces uracil.) Note that Hb C could not have originated from Hb S by change in a single base. Note further that a "compound" individual, with the Hb S gene on one

TABLE 4.1. *The Genetic Code**

Second nucleotide

Third nucleotide shown in rightmost column of each group.

First nucleotide	A or U	G or C	T or A	C or G	Third nucleotide
A or U	**AAA** *UUU* ⎫ Phe	**AGA** *UCU* ⎫	**ATA** *UAU* ⎫ Tyr	**ACA** *UGU* ⎫ Cys	A or U
	AAG *UUC* ⎭	**AGG** *UCC* ⎪ Ser	**ATG** *UAC* ⎭	**ACG** *UGC* ⎭	G or C
	AAT *UUA* ⎫ Leu	**AGT** *UCA* ⎪	**ATT** *UAA* ⎫ Stop	**ACT** *UGA* ⎱ Stop	T or A
	AAC *UUG* ⎭	**AGC** *UCG* ⎭	**ATC** *UAG* ⎭	**ACC** *UGG* ⎰ Trp	C or G
G or C	**GAA** *CUU* ⎫	**GGA** *CCU* ⎫	**GTA** *CAU* ⎫ His	**CGA** *CGU* ⎫	A or U
	GAG *CUC* ⎪ Leu	**GGG** *CCC* ⎪ Pro	**GTG** *CAC* ⎭	**CGG** *CGC* ⎪ Arg	G or C
	GAT *CUA* ⎪	**GGT** *CCA* ⎪	**GTT** *CAA* ⎫ Gln	**CGT** *CGA* ⎪	T or A
	GAC *CUG* ⎭	**GGC** *CCG* ⎭	**GTC** *CAG* ⎭	**CGC** *CGG* ⎭	C or G
T or A	**ATA** *AUU* ⎫ Ile	**TGA** *ACU* ⎫	**TTA** *AAU* ⎫ Asn	**TCA** *AGU* ⎫ Ser	A or U
	ATG *AUC* ⎪	**TGG** *ACC* ⎪ Thr	**TTG** *AAC* ⎭	**TCG** *AGC* ⎭	G or C
	ATT *AUA* ⎭	**TGT** *ACA* ⎪	**TTT** *AAA* ⎫ Lys	**TCT** *AGA* ⎫ Arg	T or A
	ATC *AUG* Met	**TGC** *ACG* ⎭	**TTC** *AAG* ⎭	**TCC** *AGG* ⎭	C or G
C or G	**CAA** *GUU* ⎫	**CGA** *GCU* ⎫	**CTA** *GAU* ⎫ Asp	**CCA** *GGU* ⎫	A or U
	CAG *GUC* ⎪ Val	**CGG** *GCC* ⎪ Ala	**CTG** *GAC* ⎭	**CCG** *GGC* ⎪ Gly	G or C
	CAT *GUA* ⎪	**CGT** *GCA* ⎪	**CTT** *GAA* ⎫ Glu	**CCT** *GGA* ⎪	T or A
	CAC *GUG* ⎭	**CGC** *GCG* ⎭	**CTC** *GAG* ⎭	**CCC** *GGG* ⎭	C or G

*The DNA codons appear in boldface type; the complementary RNA codons are in italics. A = adenine; C = cytosine; G = guanine; T = thymine; U = uridine (replaces thymine in RNA). In RNA, adenine is complementary to thymine of DNA; uridine is complementary to adenine of DNA; cytosine is complementary to guanine, and vice versa. "Stop" = punctuation. The amino acids are abbreviated as follows: Ala, alanine; Arg, arginine; Asp, aspartic acid; Asn, asparagine; Cys, cysteine; Gln, glutamine; Glu, glutamic acid; Gly, glycine; His, histidine; Ile, isoleucine; Leu, leucine; Lys, lysine; Met, methionine; Phe, phenylalanine; Pro, proline; Ser, serine; Thr, threonine; Trp, tryptophane; Tyr, tyrosine; Val, valine.

chromosome and the Hb C gene on the other homologous chromosome (the situation might be symbolized –GUA–/–AAA–), could produce a normal gamete if crossing over occurred between the appropriate adjacent bases to reconstitute the –GAA– sequence. It was first reported that in a variant hemoglobin called Hb I, aspartic acid replaced lysine as the sixteenth amino acid in the alpha chain. Francis Crick insisted that this cannot be true because change in a single base of DNA could not result in this amino acid substitution. When the matter was reinvestigated, it was found that in fact glutamic acid is substituted for lysine, a situation explicable on the basis of a single base change in DNA.

In the normal adult there is, moreover, a minor hemoglobin component called Hb A_2, that again has alpha chains that are identical chemically and genetically to those of Hb A, Hb S, and Hb F; yet another type of polypeptide chain, called the delta chain, is substituted for the beta chain.

Just as one writes the formula of water H_2O, one can write the formulas of these hemoglobins as follows:

Hb A	$= \alpha_2^A \beta_2^A$
Hb F	$= \alpha_2^A \gamma_2^F$
Hb A_2	$= \alpha_2^A \delta_2^{A_2}$
Hb S	$= \alpha_2^A \beta_2^S$
Hb C	$= \alpha_2^A \beta_2^C$
Hb Ho-2	$= \alpha_2^{Ho\text{-}2} \beta_2^A$
(Hopkins-2)	
Embryonic Hb	$= \alpha_2 \epsilon_2$

Even more specific chemical formulas for the hemoglobins stating the specific amino acid change are possible on the basis of information from fingerprinting and related analyses: for example,

$$Hb\ S = \alpha_2^A \beta_2^{6\ Val}$$
$$Hb\ C = \alpha_2^A \beta_2^{6\ Lys}$$

Up to this point, five separate loci that control the synthesis of hemoglobin have been described: (1) the alpha locus controlling synthesis of α-polypeptide chains and active throughout life, both intrauterine and postnatal; (2) the beta locus controlling the synthesis of β-polypeptide chains and active mainly in postnatal life; (3) the epsilon locus controlling the synthesis of ϵ-polypeptide chains and active only in the early embryo; (4) the gamma locus controlling the synthesis of γ-polypeptide chains and active mainly in fetal life; and (5) the delta locus controlling the synthesis of δ-polypeptide chains and active only in postnatal life. The beta, epsilon, gamma, and delta chains are each capable of combining with alpha chains to form separate species of protein: Hb A, embryonic Hb, Hb F, and Hb A_2, respectively. Figure 4.5 summarizes this scheme of the genetic control

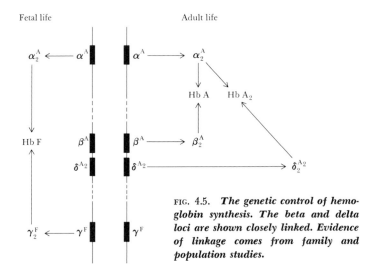

Fetal life   Adult life

FIG. 4.5. *The genetic control of hemo-globin synthesis. The beta and delta loci are shown closely linked. Evidence of linkage comes from family and population studies.*

of hemoglobin synthesis. Figure 4.6 presents the explanation of the findings when the individual is heterozygous at two of these loci. Figure 5.1 in Chap. 5 schematizes the developmental genetics of the hemoglobins.

The principle of colinearity is a leading one in biology and is the cornerstone of our understanding of much of the fine genetic structure in man. It has been proved in bacteria by Charles Yanofsky that the purine-pyrimidine bases in the DNA of a particular structural gene have the same sequence as the amino acids in the protein whose structure is determined by that gene. This means that the triplet of bases

FIG. 4.6. *Multiple hemoglobins. Persons heterozygous at both the alpha locus and the beta locus have four types of major hemoglobin. In this example, the individual is doubly heterozygous for Hb S and Hb Hopkins-2. (See page 92 and Fig. 3.15). In addition, such a person has two types of minor hemoglobin, A_2 and A_2^{Ho-2}, and in fetal life two types of fetal hemoglobin, $a_2^A\gamma^F$ and $a_2^{Ho-2}\gamma^F$ (not diagrammed here). Based on C. Baglioni, in* Molecular Genetics, *J. H. Taylor, ed. (New York: Academic Press, 1963.)*

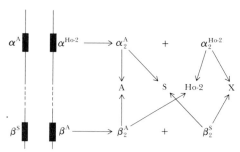

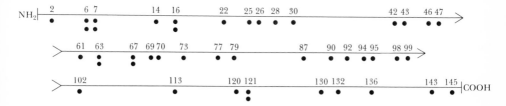

FIG. 4.7. *A mutational map of the beta hemoglobin locus, as deduced from amino acid substitutions in 44 variant hemoglobins, based on the principle of colinearity. Each dot represents an amino acid substitution found in a variant hemoglobin. From V. A. McKusick,* Mendelian Inheritance in Man *(2nd ed.), 1968, with late additions.*

which determines the amino acid at the $-NH_2$ end comes first in the gene, the triplet for the second amino acid comes second, and so on through to the last triplet, that determining the amino acid at the $-COOH$ end of the protein. In bacteria, techniques permit observation of intracistronic recombination and, therefrom, inference of the fine structure of the gene is possible. Intracistronic mapping is not possible in man. Indeed, mapping at a much grosser level is difficult (p. 32). Protein analysis is a surrogate for fine genetic analysis in man. By study of the amino acid sequence of the beta chain of normal Hb A and of a considerable number of variant hemoglobins, the map shown in simplified form in Fig. 4.7 can be derived.

The genetic control of hemoglobin synthesis is more complicated than what has been presented so far. During fetal life man has a different hemoglobin, in which the alpha chains are chemically identical to those of adult Hb A, and evidence indicates that they have the same genetic control; however, in place of the beta chain, fetal hemoglobin (Hb F) contains a chemically different *gamma* chain that is under the control of a separate genetic locus. Early in intrauterine development yet another hemoglobin, embryonic Hb, is present. This has a different nonalpha chain called *epsilon*.

In summary, then, the study of the genetics of hemoglobin synthesis and related subjects has shown that the gene (at least, many of the genes called structural genes) determines the amino acid sequence, or primary structure, of a polypeptide chain; that the secondary (helical) and tertiary (folding) structures and the physical and functional properties of the polypeptide are a consequence of its primary structure; and that more than one species of polypeptide chain, each with separate genetic control, may be combined in a single protein. (Some other proteins in man are known to be composed of two or more distinct polypeptide chains under separate genetic control.) These studies have clarified the concepts of the nature of the mutational, recombinational, and functional units of genetic material in man.

Control Mechanisms in Gene Action

Some genes have a controlling role rather than a role in determining the amino acid sequence of proteins. A possible example in man is provided again by studies of the genetics of hemoglobin synthesis. A "switch" gene seems to be involved in the change from synthesis of gamma chains (Hb F) in fetal life to the synthesis of beta chains (Hb A) in postnatal life. Persons have been found, however, with high Hb F in adult life. Incidentally, this has no apparent ill effects, even in persons who are homozygous for this mutation and have only Hb F. The regulation by the switch gene must be at the level of the chromosome because the person heterozygous for the high-F gene has about half Hb F and half Hb A. Furthermore, the person heterozygous (in repulsion) for both the high-F gene and the sickle gene has no Hb A but shows no interference with Hb S synthesis.

The homozygote for high F not only has no Hb A but also has none of the minor component normally found in the adult, Hb A_2 $(\alpha_2\delta_2)$. The delta locus (locus for δ-polypeptide chains) is closely linked with the beta locus (p. 73). Thus, a plausible model envisions an "operator" gene that controls the function of at least two structural genes, the beta locus and the delta locus (Fig. 4.8). The postulated operator gene and the beta and delta structural genes, and possibly the gamma

FIG. 4.8. (a) *A possible operon controlling the synthesis of hemoglobin.* (b) *A possible explanation of the genetic basis of hereditary persistence of fetal hemoglobin.*

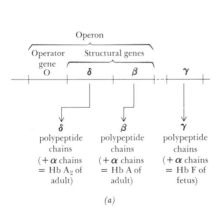

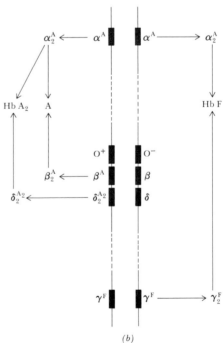

gene as well, constitute a unit of the type François Jacob and Jacques Monod have called the *operon*. In addition, there may be one or more regulator loci in other chromosomes that influence the function of individual structural genes.

Garrodian Inborn Errors of Metabolism

Up to this point, the proteins that have been considered as products of gene action are not enzymes. Hemoglobin, not an enzyme in the strict sense, has an important function in oxygen transport. The haptoglobins, transferrins, and gamma globulins are nonenzymatic serum proteins whose functions are at least partly understood. But some of the proteins specified by genes are enzymes. A mutation in the gene that determines a given enzyme may produce a disorder of the type Garrod called inborn errors of metabolism. Much has been learned about the genetic control of enzymes and about intermediary metabolism by a study of mutant forms in the human species as well as in microorganisms.

Alkaptonuria is a useful model for discussion of inborn errors of metabolism and has historical precedence, since it was the condition that was the basis for Garrod's concepts. The defect involves homo-

FIG. 4.9. *The site of the enzyme defect in three disorders of aromatic amino acid metabolism. The heavy bars indicate the block in* (a) *PKU,* (b) *albinism, and* (c) *alkaptonuria. The solid arrows indicate reactions important in normal metabolism. The broken arrows indicate reactions that become important in PKU.*

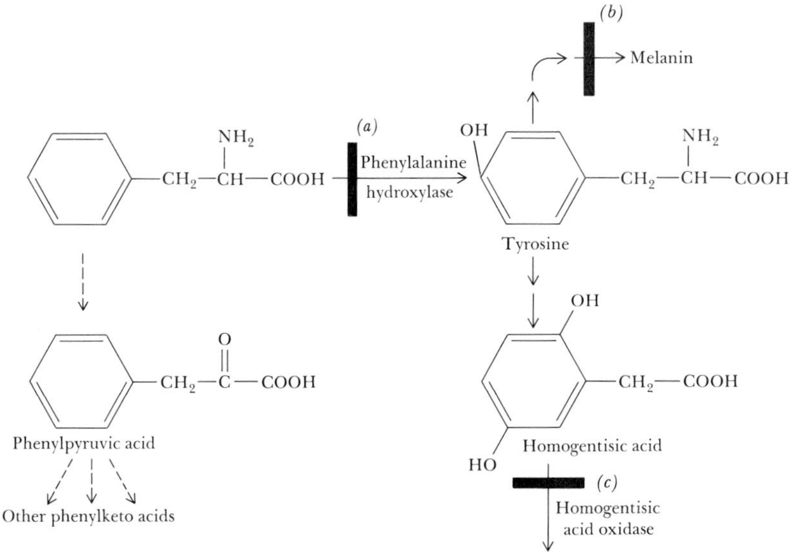

gentisic acid oxidase (homogentisicase), an enzyme that is involved in the metabolism of homogentisic acid (see Fig. 4.9c). Large amounts of homogentisic acid are excreted in the urine and turn black in alkaline urine or upon exposure to light. The black urine, causing, for example, staining of the diapers, calls attention to the condition. In addition, aggregates of homogentisic acid accumulate in the body, and become attached to the collagen of cartilage and other connective tissues. The cartilage of the ears and the sclera, which is collagenous in nature, are stained black. These manifestations are called *ochronosis*. In the joints, such as those of the spine, the accumulations lead to arthritis. Alkaptonuria is an example of a genetic enzyme block in which the phenotypic features are caused by the accumulation of excess substances just proximal to the block (Fig. 4.10b).

In some other genetic blocks in intermediary metabolism the phenotypic consequences are related to the lack of a normal product distal to the block (Fig. 4.10c). An example is albinism in which the genetic block involves a step between the amino acid tyrosine and the pigment melanin (Fig. 4.9b).

In other inborn errors of metabolism the phenotypic consequences result from excessive production of a product of what is normally an alternative and minor metabolic pathway (Fig. 4.10d). Phenylketonuria (PKU), like alkaptonuria and albinism, is a genetic defect in aromatic amino acid metabolism (see Fig. 4.9a). The defect is in the enzyme involved in the conversion of phenylalanine to tyrosine. The

FIG. 4.10. *Schematic representation of three of the ways in which a genetically determined enzyme block can produce phenotypic abnormalities:* (a) *the normal situation;* (b) *the situation in alkaptonuria;* (c) *in albinism;* (d) *in PKU.*

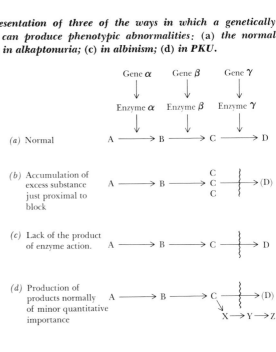

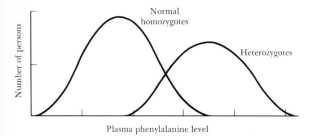

FIG. 4.11. *The phenylalanine tolerance test in the heterozygous parents of patients with PKU.*

affected person has lighter pigmentation than normal, but is not a complete albino since tyrosine is available in the diet. Severe mental retardation, one of the most prominent symptoms, is probably the result of untoward effects on brain metabolism of certain metabolic products of phenylalanine formed through alternative pathways. Certain of these alternative metabolites of phenylalanine, especially phenylpyruvic acid, are excreted in the urine and are one basis for diagnosis of the disorder. The difference in phenotype of these three diseases—alkaptonuria, PKU, and albinism—despite the fact that they involve closely related metabolic steps, is noteworthy.

Essentially all inborn errors of metabolism are inherited as recessives; the clinical disorder is present only in the homozygote. The heterozygote does not manifest the disorder; apparently a double dose of

FIG. 4.12. *Level of galactose-1-phosphate uridyl transferase in the erythrocytes of normals, of homozygous galactosemic subjects, and of heterozygous parents of galactosemic children. Redrawn from H. N. Kirkman and E. Bynum,* **Ann. Human Genet., 23 (1959), 117.**

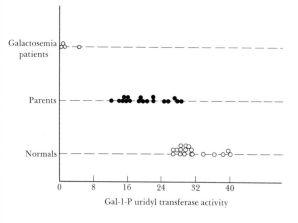

enzyme is no better than a single dose under usual circumstances. In several conditions, however, the heterozygous carrier can be identified by special means, one of which involves stressing the particular enzymatic step. In PKU, for example, the phenotypically normal but genetically heterozygous parents of affected persons tend to show blood levels of phenylalanine that are higher and last longer than normal when a standard dose of this amino acid is administered. This is the so-called phenylalanine tolerance test (Fig. 4.11).

A second method for demonstrating the heterozygote is illustrated by galactosemia. In this disorder, absence of the enzyme galactose-1-phosphate uridyl transferase (gal-1-P uridyl transferase) renders the homozygote incapable of metabolizing galactose of milk. The enzyme defect can conveniently be demonstrated in the circulating erythrocytes. The heterozygote tends to have an enzyme level intermediate between that of the two homozygotes, the normal and the affected (Fig. 4.12).

Defects in Active Transport Mechanisms

Garrod investigated four conditions he considered inborn errors of metabolism: alkaptonuria, albinism, cystinuria, and pentosuria. It is now known that one of these, cystinuria, is not really an inborn error of metabolism. Large amounts of cystine appear in the urine not because of a defect in intermediary metabolism but rather because of a defect in the renal tubule mechanism by which cystine is resorbed from the glomerular filtrate. Normal resorption is accomplished by an active transport mechanism; that is, a metabolic process is involved in transferring cystine and other amino acids from the lumen of the renal tubule (where it has arrived by filtration through the glomerulus) to the blood stream on the other side of the renal tubule cell. At the molecular level the distinction between genetic defects in active transport systems and inborn errors of metabolism is probably artificial; enzyme defects may be involved in both. Other genetic defects in active transport mechanisms have been discovered.

Dosage Effect

In a number of instances in which it is possible to measure the primary product of gene action, one finds that the heterozygote has about half as much of the product as does the homozygote. The example of gal-1-P uridyl transferase in galactosemia heterozygotes has already been cited in Fig. 4.12. Sometimes the heterozygote does not have exactly half as much product protein as the homozygote. For example, the sickling heterozygote, SA, has normal levels of total hemoglobin, but Hb A represents about 70 percent rather than 50 percent of the total. (This is evident in Fig. 4.3.) The reason for the inequality is currently under study.

Dosage Effect in Connection with X-linked Genes

For genes on the X chromosome a special problem of dosage effect exists, and the following questions arise: Does the female with two X chromosomes have twice as much gene product, such as the enzyme G6PD or the blood-clotting factor antihemophilic globulin, as does the male with one X chromosome? If a dosage effect of the two X chromosomes is not observed in the normal female, what is the mechanism of dosage compensation?

One could imagine that to avoid disruptive dosage effects the X chromosome might have been largely stripped of genetic information in evolution. That this is not the case, however, is indicated by the considerable list of traits known to be determined by genes on the X chromosome. The X chromosome seems to carry at least as much genetic information as an autosome of comparable length. Furthermore, studies suggest that the normal male and normal female have essentially identical amounts of gene product for a number of traits determined by genes on the X chromosome.

The Lyon hypothesis (Fig. 4.13) provides an explanation for both

FIG. 4.13. *A schematic representation of the Lyon hypothesis.*

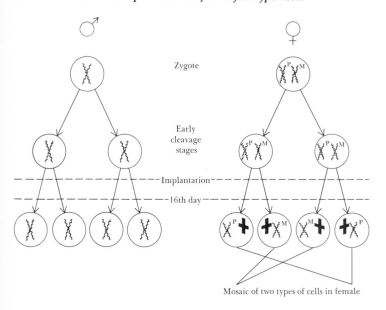

Mosaic of two types of cells in female

dosage compensation and for the findings in heterozygous females. As outlined earlier (p. 20), the explanation that occurred to Mary Lyon of Harwell, England, and simultaneously to several other workers, is that after a time early in embryogenesis one X chromosome becomes genetically inactive and forms the Barr body of interphase nuclei (p. 13). The Lyon hypothesis further suggests (1) that it is a random matter as to which X chromosome in any single cell—whether the one derived from the father or the one from the mother—is the inactive one, and (2) that once the differentiation of the X chromosome has occurred in a given cell, with one X chromosome assuming an inactive role, then the *same* X chromosome remains inactive in all descendants of that cell. The primordial germ cells of the female, even though they have two X chromosomes, do not participate in this process of X-chromosome differentiation.

The Lyon hypothesis provides a mechanism for dosage compensation since the female has no more *active* X chromosomes than does the male. The Lyon hypothesis is obviously not in conflict with the facts of X-linked recessive inheritance. In the hemizygous male whose X chromosome carries, for example, the mutant gene for hemophilia, every gene in his body at the locus for antihemophilic globulin (which is deficient in hemophilia) is of the mutant type. In the heterozygous female who carries the hemophilia gene on one X chromosome and the wild-type allele on the other X chromosome, half the antihemophilic-globulin genes, on the average, are of the mutant type and half are wild type, if there is an equal probability that the mutant X chromosome will be the active or the inactive one in any given body cell.

The Lyon hypothesis also provides an explanation for the intermediate level of gene product in the heterozygous female and for the rather wide variability in the level of gene product observed in heterozygotes of several X-linked disorders. Since the decision as to which X chromosome will be the inactive one is made early in embryogenesis, the number of cells is relatively small. Especially small is the number of pertinent anlage cells destined for a particular function, let us say, synthesis of antihemophilic globulin; perhaps only about a dozen such cells are present at the "time of decision." By chance alone, rare individuals might have all cells with the mutant X chromosome as the active one; such individuals would be hemophiliacs—so-called manifesting heterozygotes. Or all cells might by chance have the wild-type X chromosome active; these individuals would have normal levels of antihemophilic globulin. But the great majority of heterozygous females would have an intermediate proportion of cells with the mutant X chromosome active.

If the Lyon hypothesis is valid and all cells in a particular line have the same X chromosome inactive, one would anticipate that the heterozygous female would display mosaicism for those X-linked traits in which the phenotype is evident at the cellular level. Mosaicism of the predicted type has been demonstrated for several conditions in heter-

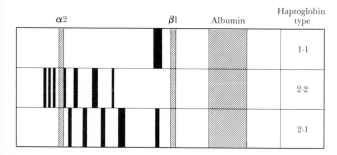

ozygotes; notable among the examples is G6PD deficiency. Heterozygous females have been shown to have two populations of cells, those with normal enzyme activity and those with very little enzyme activity.

Although the Lyon hypothesis cannot be considered completely proved, the evidence to date suggests that it is substantially correct and that dosage compensation for X-borne genes is determined mainly by this mechanism.

Gene Interaction

Gene interaction is observed between alleles and between non-alleles. Allelic interaction is well illustrated by the example of haptoglobin, a serum protein that shows genetic variation (Fig. 4.14). Two alleles that occur frequently have been recognized—haptoglobin-1 (Hp^1) and haptoglobin-2 (Hp^2). The haptoglobin types of the two homozygotes are referred to as Hp 1-1 and Hp 2-2. In the heterozygote, the haptoglobin is not a mixture of Hp 1-1 and Hp 2-2 proteins; instead the heterozygote has a unique protein product with chemical properties quite distinct from those of the proteins produced by the two homozygotes. It has been suggested that some instances of heterozygote advantage might have a basis in the formation of a unique protein by the heterozygote through gene interaction.

Nonallelic interaction is illustrated by several phenomena associated with blood groups, and is best understood by a discussion of them.

Immunogenetics: Blood Groups

The blood groups share with the hemoglobins the distinction of having contributed heavily to the formulation of principles of human genetics and of genetics in general. Blood groups are, furthermore, of great significance in medicine. It would be well for the student to become acquainted with a few basic facts about the blood groups and to understand their importance to human genetics and to medicine.

The blood groups are genetically determined antigens of the erythrocytes. At least 15 different blood-group systems, each determined by a separate locus, have been identified. Furthermore, at most of these separate loci multiple alleles are now known. The various blood group systems which are polymorphic include, in order of discovery, ABO, MN, P, Rh, Lutheran, Kell, Duffy, Kidd, Lewis, Diego, Yt, Auberger, Xg, Dombrock, and Stoltzfus. In addition to these systems, each with two or more common alleles, a considerable number of "private" blood groups (very rare red cell antigens) and "public" blood groups (antigens carried by very nearly everyone) are known.

The different antigens on the red cells are identified by means of antibodies—proteins in serum that combine with the antigens and produce such effects as agglutination of these cells. Some of the antibodies occur naturally; for example, in the ABO blood-group system, type A persons have in their serum antibody against type B cells; type B persons have antibody against type A cells; and type O persons have antibody against both type A and type B cells. Consequently Landsteiner was able to demonstrate the ABO blood types by mixing the serum and red cells of different persons.

To demonstrate the antigens of other blood groups, it is necessary to obtain the corresponding antibody by one of two methods. In the case of the MN blood groups the antibody was produced in another species, the rabbit, by injecting human red blood cells into rabbits. Rabbits injected with cells from persons of the MM genotype produced anti-M serum; cells from persons of the NN genotype stimulated production of anti-N serum. Cells from persons of the MN genotype, that is, heterozygotes, were agglutinated by either anti-M or anti-N rabbit serum.

The other method by which antibodies demonstrating blood groups are developed is the accidental development of antibody by a pregnant woman or a transfused patient. If the mother lacks a red cell antigen that is present in the fetus (who inherited it from the father), the mother may develop antibodies against the antigen when fetal cells leak over into the mother. The Rh blood types, as well as several of the others, were discovered largely through this mechanism. A situation that in principle is exactly the same occurs when a patient, transfused with red cells containing an antigen he does not possess, develops a serum antibody against that antigen. The X-linked blood group Xg[a] was discovered in this way.

Blood groups are of great genetic significance; they provided some of the clearest early examples of simple Mendelian inheritance. Most blood groups are codominant. In many instances—in fact so often that it is now considered a general principle—an antibody is eventually discovered for both antigens present in the heterozygote. Thus, in the Kell system an antiserum was first discovered that agglutinated the erythrocytes from persons of the genotype *KK* or *Kk* but not of persons of genotype *kk*. Later an antiserum was found that agglutinated the cells from persons of the genotype *Kk* or *kk*. ("Codominant" is not to be confused with "partial dominant," or "incomplete dominant," terms used when the phenotype of the homozygote is different from that of the heterozygote.)

The Hardy-Weinberg principle (p. 160) was first put to test in connection with the ABO blood groups. Multiple allelism was first demonstrated in man in the case of the ABO types. In fact, Bernstein proved multiple allelism (rather than two loci, one determining A and non-A and the other B and non-B) by showing that the phenotype frequencies agree with those predicted by the Hardy-Weinberg principle for a multiple-allele system.

Most of the markers useful for genetic linkage studies are blood groups. Population genetics has made extensive use of the blood groups; the influence of selection, drift, and gene flow has been studied using these groups. Blood groups also illustrate the difficulties in distinguishing close linkage and allelism in man. A transatlantic polemic raged over this matter in regard to the Rh blood-group system. Sir Ronald Fisher and Robert R. Race in England suggested that three closely linked loci were responsible for the Rh specificities that they termed C, D, and E. Alexander Wiener in this country insisted that a single complex locus was involved.

The secretor and ABO loci show nonallelic interaction. The secretor trait is characterized by the secretion of ABO blood group antigen in the saliva. About 70 percent of persons of northern European origin are secretors. But to secrete type A antigen the person must be both of genotypes *AA, AB,* or *AO* and of genotype *SS* or *Ss*. And to secrete type B antigen the person must be both of genotype *BB, AB,* or *BO* and of genotype *SS* or *Ss*. Imagine that nothing was known of erythrocyte

FIG. 4.15 *The Bombay phenomenon in an Italian-American family. The grandparents, I 1 and I 2, were first cousins. After P. Levine et al.,* **Blood, 10** *(1955), 1100.*

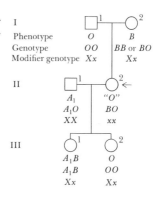

types and that the only phenotype that could be tested was the presence or absence of A antigen in the saliva. The distribution of these phenotypes in families and in populations would be quite different from that of phenotypes dependent on a single gene difference.

Suppression is illustrated by the rare Bombay phenomenon—so called because it was first detected in a family in India. Persons were found who were of blood group O but who, from the blood types of their parents and children (Fig. 4.15), were known to carry the gene for blood type B (or A). The explanation assigned to the findings was that these individuals are homozygous for a rare recessive suppressor gene (*xx*), or viewed differently, for a mutation in a gene essential to the development of the ABO blood-group antigen. In the absence of at least one dominant *X* gene a precursor substance may not be formed.

The importance of blood groups to medicine lies in at least three areas. (1) Blood transfusion with no resulting complications requires recognition and understanding of these genetic differences. (2) The proper management and prevention of the ill-effects of materno-fetal incompatibility (for example, Rh problems of pregnancy) likewise

FIG. 4.16. *Complicated paternity.* (a) *Alleged paternity with blood groups. The blood groups: CDE, Rh types; Lu, Lutheran types; K, Kell types; Le, Lewis types; Fy, Duffy types.* (b) *Interpretation of paternity on basis of blood groups. Redrawn from Race and Sanger (1968).*

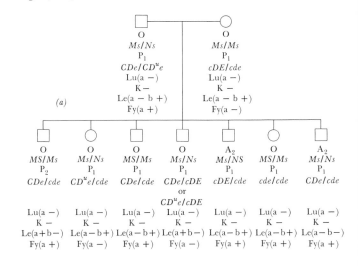

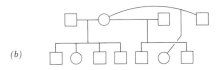

must be based on a clear understanding of blood group principles. (3) Medicolegal applications include cases of disputed parentage. Usually there is no question about the identity of the mother and only paternity is in doubt. Paternity can never be proved with absolute certainty but can be disproved in two ways: (a) A man is excluded as the father if he and the mother both lack an antigen that the child possesses. The child cannot have an antigen lacking in both parents— barring rare phenomena such as the Bombay trait and mutation. (b) A man is excluded if the child fails to show an antigen he must transmit. For example, an AB man cannot have an O child, nor can an M man have an N child.

The use of blood groups to exclude paternity is illustrated by the example shown in Fig. 4.16: in *a* are shown the blood groups of the parents and seven children purported to be one family (it is clear that the alleged father cannot have sired the last three children) ; in *b* is shown what Race and Sanger (1968) referred to as "the most likely, and the most charitable, interpretation."

Immunogenetics: Histocompatibility and Organ Transplantation

Transfusion is a form of tissue transplantation and in general the rules of histocompatibility involved in transplantation of kidneys and other organs are the same as those already outlined for blood groups. The blood groups and the antigenic components of the walls of cells in other tissues are similar; indeed some of the erythrocyte groups are important in histocompatibility. For example, the ABO blood groups represent a strong "histocompatibility barrier." The donor of an organ for transplantation should not have an ABO type not present in the recipient. Thus, an AB recipient can accept (assuming that other types discussed later are compatible) a transplant from an A, B, or O donor as well as from another AB person, but a type O recipient can accept a transplant from none of these except another type O person. Histocompatibility antigens, like blood groups, are codominant in their inheritance.

The blood group antigens and histocompatibility antigens certainly do not have as their sole function the embarrassment of the transplantation surgeon or the scientific delight of the geneticist, but rather are important constituents of the cell wall with specific structural and functional roles. Their antigenic roles may have selective significance. They might, for example, protect against malignancy by leading to the elimination of cells which have suffered a somatic mutation with malignant potential. They might also protect against foreign proteins which come from other humans and are carried into a host by a virus.

In addition to the ABO blood group locus, the important histocompatibility locus in man is that which is called HL-A, because it is studied by means of antisera which type *h*uman *l*eukocytes (the A meaning that this is the first human leukocyte locus to be designated). The antisera are derived from women who have had many pregnancies and have become sensitized to leukocytes or other tissue cells of their offspring, or from persons who for some medical reason have received many transfusions. Cytotoxic effects of the antiserum, that is, abnormalities observed in leukocytes under the microscope after exposure to the antiserum, are the indication of a positive reaction. Studies indicate that a very large number of alleles, *several dozen at least,* exist at the HL-A locus. It is thought to be homologous to the H-2 locus, the major histocompatibility locus of the laboratory mouse. In both, as in the case of the Rh locus also, it is uncertain whether a closely linked cluster of genes or a single locus is involved, but the question is perhaps of theoretical interest only.

The large number of alleles at the HL-A locus means that compatibility is more likely to be found among sibs than between parent and offspring and that matching of unrelated persons is difficult. Because of the large number of alleles, a high proportion of persons are heterozygous. Furthermore, two parents have a strong probability of being heterozygous for different alleles. Thus, the father might have alleles A and B and the mother alleles C and D. The combinations among the offspring would, then, be AC, BC, AD, BD and none of the children would cross-match with either parent. On the other hand, sibs would stand a 25 percent chance of being compatible. It can be shown that as the number of alleles at a histocompatibility locus increases, the proportion of unrelated individuals who are compatible approaches zero, but the proportion of compatible sibs is never less than 25 percent.

Other, weaker histocompatibility loci are not tested for by the HL-A system. Even sibs compatible for ABO blood type and for HL-A type will reject a skin graft, for example, although rejection is delayed. Skin grafting is used in distinguishing monozygotic and dizygotic twins (see p. 118). In transplantation of organs, particularly kidneys, the best possible match is sought. Even if the donor and recipient are of the same ABO and HL-A types, it is necessary to give the recipient certain drugs which suppress the immunologic rejection process.

Another test for histocompatibility is the MLC (mixed leukocyte culture) test. Lymphocytes from incompatible persons, when mixed in the test tube, mutually stimulate each other, with initiation of DNA synthesis and mitosis in a manner similar to the nonspecific effects of phytohemagglutinin (p. 9). Matching, or pairing, can be done by this method. Treatment of the lymphocytes from one person with mitomycin C, which blocks the mitogenic effect on that cell, makes it a one-way test whose results are easier to interpret. The MLC test and leukocyte typing by means of antisera give results which are consistent with each other.

References

Baglioni, Corrado, "Correlations Between Genetics and Chemistry of Human Hemoglobins," in *Molecular Genetics,* Vol. I, J. H. Taylor, ed. New York: Academic Press, Inc., 1963, pp. 405–75.

Dayhoff, M. O., and R. V. Eck, *Atlas of Protein Sequence and Structure 1967–68.* Silver Spring, Md.: National Biomedical Research Foundation, 1968.

Garrod, Archibald E., *Inborn Errors of Metabolism.* Reprinted with a supplement by Harry Harris. New York: Oxford University Press, 1963.

Giblett, Eloise R., *Genetic Markers in Human Blood.* Oxford, England: Blackwell Scientific Publications, 1969.

Ingram, Vernon M., *Hemoglobins in Genetics and Evolution.* New York: Columbia University Press, 1963.

Perutz, M. F., and H. Lehmann, "Molecular Pathology of Human Haemoglobin," *Nature, 219* (1968), 902–09. An exciting review of mutations in hemoglobin and the relation between change in structure and change in function.

Race, Robert R., and Ruth Sanger, *Blood Groups in Man,* 5th ed. Oxford, England: Blackwell Scientific Publications, 1968.

Stanbury, J. B., J. B. Wyngaarten, and D. S. Fredrickson, eds., *The Metabolic Basis of Inherited Disease.* New York: McGraw-Hill Book Company, 1960. A comprehensive survey of inborn errors of metabolism.

$$\mathcal{F}ive$$

Genes in Development
and Differentiation

The central question of embryogenesis and differentiation is why certain genes function only at certain times and in certain tissues. All somatic cells contain the same complement of genes, yet only erythroblasts synthesize hemoglobin and only liver cells synthesize serum albumin, to cite two examples.

Hemoglobin provides a particularly clear example in man of the function of one gene in fetal life and another in extrauterine life. The genetic locus responsible for production of α-polypeptide chains of hemoglobin is functional throughout the life of the organism, beginning at a relatively early stage of embryogenesis. However, the separate genetic locus determining production of β-polypeptide chains of hemoglobin is quiescent in fetal life, when a third genetic locus is active and determines the production of γ-polypeptide chains that combine with α-polypeptide chains to form fetal hemoglobin. At about the time of birth the gamma locus is "turned off" and the beta locus is "turned on" (see Fig. 5.1). An operator gene (Fig. 4.8) appears to be involved in turning on the beta locus, but the problem is only moved back one step to the question of what controls the operator gene.

Another striking example of gene action limited to one phase in the life of the human organism, and probably one tissue, is placental alkaline phosphatase. This specific enzyme is determined by a gene apparently active during

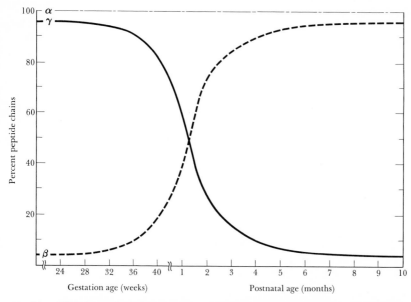

FIG. 5.1. *Shift from fetal hemoglobin to adult hemoglobins (**Hb A** and **Hb A$_2$**) in the perinatal period. In earlier stages of intrauterine development, the nonalpha chain of hemoglobin is different from the gamma and beta chains, called epsilon (ϵ), and the hemoglobin ($a_2\epsilon_2$) is called embryonic Hb. A shift from espsilon chain synthesis to gamma chain synthesis occurs in the first part of pregnancy.*

fetal life only—a gene of the fetus since the placenta of man is largely fetal in origin. It is likely that some Mendelizing congenital malformations are the result of defects in a protein whose synthesis occurs only during a restricted period of early development. In such instances one would not expect to be able to identify a fundamental chemical abnormality in the adult.

Studies of the enzyme lactic acid dehydrogenase (LDH) provide evidence of differences in genic activity in different tissues. This enzyme protein is a tetramer, consisting of four polypeptide chains. The chains can be of two different types, A and B. Five different proteins are formed depending on the proportions of A and B chains: A_4, A_3B_1, A_2B_2, A_1B_3, and B_4, the last having most rapid electrophoretic migration. (These are sometimes referred to as LDH 5, 4, 3, 2, and 1, respectively.) Each of these five proteins has characteristic physical and enzymatic properties, and which ones of the five proteins are present in a given tissue is a characteristic of that tissue at a particular stage of development. During development the proportions of the five enzymes change in many tissues. For example, all human embryonic tissues have predominantly B_4 (LDH 1), as does adult heart muscle, but adult skeletal muscle has predominantly A_4 (LDH 5). The A and B chains are apparently under separate genetic control; a mutation involving each has been found in man. The relative activity of these genes determines which of the five enzymes predominate in a given cell at a given stage.

Intrauterine selection is undoubtedly rigorous, even under the presently improved conditions of prenatal care in prosperous countries. It is estimated that 15 to 25 percent of zygotes are lost before birth. Many of these may have gross chromosomal aberrations—such as triploidy, which has been identified in some abortions (see p. 30). Others probably have more subtle defects. Materno-fetal incompatibility for the ABO blood groups is known to lead to early fetal death and abortion. It has been estimated that as many as 5 percent of conceptions are lost through ABO incompatibility. This phenomenon is not surprising, since the type O mother, for example, has "natural" antibody against type A and B antigens of fetal erythrocytes and other tissues, including perhaps the placenta, which is largely of fetal genotype.

In the Rhesus (Rh) system, unlike the ABO blood-group system, the mother does not naturally carry antibodies against the types of Rhesus antigens that she does not possess. Trouble develops only if the mother becomes sensitized to what for her is a foreign antigen entering her system in the form of red cells from the fetus that contains a different Rhesus antigen through inheritance from the father. Sensitization of the mother is accomplished by actual bleeding from the fetus into the mother. Traumatic delivery is particularly likely to cause entry of fetal red cells into the maternal circulation. Sensitive methods based on the demonstration of red cells containing fetal hemoglobin can detect the presence of as little as 1 cc of fetal blood distributed in the mother's circulation. In subsequent pregnancies, if the fetus is again incompatible, the mother's titer of anti-Rh antibody may rise briskly, and the damaging effect on the infant may result in the characteristic clinical picture of erythroblastosis fetalis. See Fig. 5.2.

About 15 percent of Caucasians are Rhesus negative. Erythroblastosis fetalis would be much more frequent than it is if no other factors were involved. One factor is undoubtedly the occurrence of feto-maternal bleeds; many Rhesus-incompatible pregnancies pass without the occurrence of such sensitizing bleeds. Another important factor protecting against Rhesus sensitization is accompanying ABO incompatibility. If the erythrocytes that bleed from the fetus to the mother are of a different ABO blood type than the mother's, the natural antibody of the mother is likely to knock them out before they can incite sensitization.

The main offender in the Rhesus type of erythroblastosis fetalis is antigen *D*. It seems to have greater antigenic propensities than the antigens *C* and *E*. Obviously, once erythroblastosis has occurred, the risk that future children of an Rh-negative mother and an Rh-positive

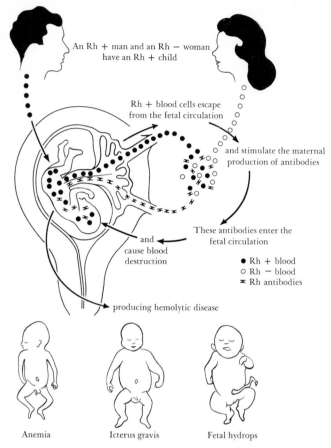

An Rh + man and an Rh − woman have an Rh + child

Rh + blood cells escape from the fetal circulation

and stimulate the maternal production of antibodies

These antibodies enter the fetal circulation

and cause blood destruction

● Rh + blood
○ Rh − blood
✖ Rh antibodies

producing hemolytic disease

Anemia   Icterus gravis   Fetal hydrops

FIG. 5.2. *The mechanism of* erythroblastosis fetalis. *Redrawn from E. L. Potter,* **Rh: Its Relation to Congenital Hemolytic Disease and to Intragroup Transfusion Reactions (*Year Book, 1947*).**

father will have erythroblastosis is 100 percent if the father is *DD* but is only 50 percent if he is *Dd*. Since in various populations some *CDE* combinations are more frequent than others, a probabilistic estimate of whether the father is *DD* or *Dd* can be obtained from the reactions with anti-*C,* anti-*c,* and anti-*E* antiserums. (Unfortunately no anti-*d* antiserum is available.) For example, if the father is an Englishman and his erythrocytes react to anti-*C,* anti-*D,* and anti-*e,* but not to anti-*c* and anti-*E,* he must have *CDe* on one chromosome. The other chromosome may carry either *CDe* or *Cde.* Since in English populations *CDe* is about 41 times more common than *Cde,* it is likely that the man is homozygous *DD* (that is, *CDe/CDe*) .

Prevention of sensitization of Rh negative women who bear offspring by Rh positive men is now possible. The passage of fetal red cells into the maternal circulation ("transplacental bleed") is recognized by staining the mother's blood smear for red cells containing fetal hemoglobin. Most transplacental bleeds occur at the time of delivery. If the baby is Rh positive and if the transplacental bleed is of significant volume, anti-*D* gamma globulin is administered intravenously to the mother after delivery of the baby. It has been shown that the agent clears the fetal cells from the mother's blood and prevents sensitization.

The Sex Ratio

The ratio of males to females observed at birth is called the secondary sex ratio and is about 1.06 in whites in the United States. The primary sex ratio, that of the zygotes, is considerably higher, perhaps 1.30. The reason that Y-bearing sperm have an advantage over the X-bearing sperm in fertilization is unknown. Whatever the explanation, sex chromatin studies of early embryos show an excess of males in a ratio of about 1.3 to females. What accounts for the decline in the ratio between conception and birth? One possibility is that a proportion of male zygotes succumb to X-linked recessive lethals, but there is little direct evidence to support this plausible explanation. Another possibility is that the female with two X chromosomes enjoys heterotic vigor. Certainly the female has a greater survival throughout the span of life, both intra- and extrauterine. The two sexes become numerically equal in middle life and in the older age groups women are more numerous than men.

Twins

About 1 percent of all pregnancies in whites in the United States are twin pregnancies (counting only those terminating in live births). Therefore about 2 percent of all newborn babies are twins. Since twins have a somewhat reduced chance of surviving the first year of life as compared to nontwins, the frequency of twins at age one year and older is about 1.9 percent.

It is well known that twins are of two types. Dizygotic, two-egg, or fraternal twins result when two ova are produced at about the same time and both are fertilized. Monozygotic, one-egg, or identical twins result from the splitting of the zygote at an early stage. Monozygotic twins are of course genetically identical, barring somatic mutation. Dizygotic twins are genetically no more similar than ordinary sibs.

Monozygotic twins are of course either both male or both female. Dizygotic twins are of like sex or of unlike sex in approximately equal frequency. All twins of unlike sex are dizygotic, but twins of like sex may be of either type.

In 1901 Weinberg, a physician in Stuttgart, suggested a method of predicting the proportion of like-sexed twins that are monozygotic. Dizygotic twins have an about equal probability of being of like or of unlike sex. If n is the number of twins of unlike sex in a randomly ascertained series, then that same number n of the twins of like sex are also dizygotic; the remainder of the twins of like sex are monozygotic. In this country, for example, about 31 percent of white twins are of unlike sex and therefore dizygotic. The remaining 69 percent, the group of twins of like sex, includes another 31 percent that are dizygotic. The remaining 38 percent are monozygotic.

Monozygotic twinning shows almost no effect from the mother's age. Dizygotic twinning, on the other hand, is more frequent in older mothers. The frequency of twinning varies in different ethnic stocks, even when allowance is made for variations in the average age of mothers. In the 1950s whites in the United States had about 10 twin pregnancies per 1,000, and nonwhites (mainly Negroes) had over 13.5 twin pregnancies per 1,000. Almost all the excess twin pregnancies in the nonwhites were dizygotic. The racial differences in the frequency of dizygotic twinning may be evidence for genetic factors. That twinning shows a familial aggregation is further evidence of genetic factors. The propensity for dizygotic twinning is, of course, expressed only in females; females from high twinning families have an increased chance of bearing dizygotic twins. Males of such families may transmit this propensity to their daughters, although they do not sire dizygotic twins with increased frequency.

Environmental factors probably also influence the rate of twinning, since the same ethnic stocks in different habitats may show different rates. Furthermore, there was a significant decline in the rate of twinning in white people in the United States in the period from 1922 to 1958. In Sweden there has been a decline in twinning over the last two centuries.

Monozygotic twins can be anyone of four types: (1) the zygote may be divided at first cleavage; (2) two inner cell masses may develop (Fig. 5.3*a*); (3) a single cell mass may be formed which subsequently divides (Fig. 5.3*b*); or (4) the division may be particularly late and incomplete, to result in conjoined twins ("Siamese twins"). That twinning may occur at the first cleavage is indicated by observation of twins who are identical as indicated by mutual acceptance of skin grafts but differ in sex chromosome constitution, one being XO, and the other XY. (Here is an exception to the rule that twins of unlike sex are dizygotic!) The placenta in such cases is likely to be of type *a* or type *b* (Fig. 5.4). If two cell masses develop, the placenta is of type *c*. If one cell mass undergoes splitting the placenta is of type *d*.

The twin method was proposed by Francis Galton in 1876 as a means

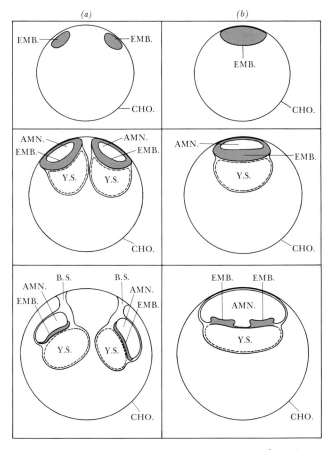

FIG. 5.3. *Two types of monozygotic twinning in man:* (a) *by formation of two inner cell masses;* (b) *by formation of two embryos on a single germ disc. From* **G. W. Corner,** Bull. Johns Hopkins Hosp., 33 *(1923) 389.*

of distinguishing between environment and heredity, nurture and nature, in the determination of human variation. The concordance method (see p. 190) compares the degree of similarity of monozygotic twins to that of dizygotic twins of like sex, and compares the similarity of monozygotic twins that have been reared together to those that have been reared apart. The method obviously requires precise diagnosis of zygosity. Two methods of zygosity diagnosis are available: examination of the placenta and the similarity method, including skin grafting.

The developing zygote invests itself in two membranes, an inner (the amnion) and an outer (the chorion). In the case of dizygotic twins each twin has completely separate membranes, although when implantation is close together the two placentas may fuse. A significant proportion of monozygotic twins also have two chorions and two amnions, and some have two placentas. (See Fig. 5.4.) All monochorionic twins are monozygotic, and about 70 percent of monozygotic twins have one chorion. Only in these cases is an unequivocal diagnosis of monozygosity possible on the basis of placental findings. The four types of placentation shown in Fig. 5.4 are as follows: (*a*) diamniotic dichorionic separate; (*b*) diamniotic dichorionic fused; (*c*) diamniotic monochorionic; and (*d*) monoamniotic monochorionic.

Table 5.1 presents data on the frequency of the four placental types in two series. Placentation permitted definitive diagnosis of monozygosity in 31 percent (77/250) and 22 percent (118/549), respectively, in these series. By Weinberg's method, it can be deduced that in Benirschke's series 25 of class *b* twins were monozygotic and 8 of class *a* twins were monozygotic. The estimated total of monozygotic twins is, therefore, 110 or 44 percent of the series. The placentation permitted definitive diagnosis of monozygosity in 77 of the 110, or 70 percent. In 70 cases of 250 (28 percent) the unlike sex of the twins established dizygosity. Thus, definitive zygosity was evident at birth in 59 percent of the second series of twin pairs.

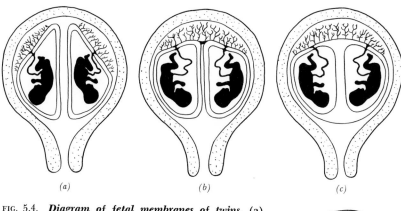

(a)   (b)   (c)

FIG. 5.4. *Diagram of fetal membranes of twins.* (a) *Monozygotic or dizygotic twins with separate amnions, chorions, and placentas.* (b) *Monozygotic or dizygotic twins with separate amnions and chorions but single placenta.* (c) *Monozygotic twins with separate amnions but single chorion and placenta.* (d) *Monozygotic twins with single amnion, chorion, and placenta. Redrawn from E. L. Potter,* Fundamentals of Human Reproduction *(New York: McGraw-Hill Book Co., 1948).*

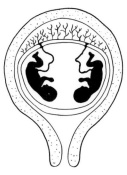

(d)

TABLE 5.1. *Placentation in Two Series of Twins*

Placental type (zygosity)	Benirschke (1961)		Potter (1963)	
	Number of placental type	Like sex	Number of placental type	Like sex
a. Diamniotic dichorionic separate (MZ, DZ)	88	48	239	141
b. Diamniotic dichorionic fused (MZ, DZ)	85	55	192	113
c. Diamniotic monochorionic (MZ)	74	74	117	117
d. Monoamniotic monochorionic (MZ)	3	3	1	1
TOTAL	250	180	549*	372

* In 19 additional cases the placenta was probably single and monozygotic.

The similarity method is usually the only one available for zygosity diagnosis and is more reliable than placental examination. However, as in many other situations in human genetics the conclusions are based on probability. Although in many cases monozygosity can be excluded with certainty, it can be established only with a certain *probability* (see p. 149), and never with absolute certainty. Objective single-gene traits such as blood groups and serum protein groups are used. The objective markers are the same as those used for linkage studies (see Chap. 3). Morphologic characteristics such as eye color, nose configuration, ear form, and others are less reliable. Ridge counts, a quantitative evaluation of the fingerprints (see p. 158) under polygenic control, can be used in zygosity diagnosis. Methods of mathematically estimating the probability that a given set of twins is monozygotic are reviewed on p. 149.

The last court of appeal in zygosity diagnosis is skin grafting. Reciprocal grafts should, of course, "take" in monozygotic twins and will not "take" in dizygotic twins. A skin graft from anyone other than a monozygotic twin is rejected eventually, although the time for rejection is variable. It is doubtful that vascular communications in dizygotic twins often lead to immune tolerance so that reciprocal skin grafts would be successful. However, there is no extensive experience with skin grafting in twins.

In cattle, the female of a pair of twins of unlike sex may develop into what is called a *freemartin*. The external genitalia are female in type, but the ovaries do not develop normally and the animal is sterile.

A freemartin is formed only if fusion of fetal membranes takes place leading to cross-circulation. Hormones from the male were thought to be responsible for the changes in the female cotwin. However, recent work demonstrates actual germ-cell chimerism. No exactly comparable situation is known in man.

It is certain that vascular exchange between dizygotic human twins occurs, however, since chimerism of blood type has been recognized. Several examples have been found in which a dizygotic twin had a minor population of erythrocytes with a blood group different from that of the majority. Primordial blood cells from the cotwin colonized the chimera twin in fetal life. In all instances tested the chimera twin could be shown to have developed immune tolerance and would accept a skin graft from the other twin. Sometimes both, and sometimes only one, of a pair of twins showed chimerism. Precursors of leukocytes may become grafted into a dizygotic cotwin by the same mechanism. Male twins have been found who had female cotwins and who showed polymorphonuclear leukocytes with the "drumstick" (p. 14) characteristic of the female. In some instances, study of the chromosomes of leukocytes has shown the karyotype of the opposite sex, that is, some XY cells in a female cotwin.

Congenital Malformations ("Birth Defects")

The term *congenital* is not synonymous with the term *hereditary*. *Congenital* means "present at birth." Malformations evident at birth and therefore congenital may be overwhelmingly genetic in their cause, or, on the contrary, may be overwhelmingly determined by extrinsic factors (see p. 181). Furthermore, not all genetic disorders are congenital—at least in terms of the phenotypic change being evident at birth. For example, in Huntington's chorea the fundamental defect is presumably present in the nervous system at birth, but the phenotypic manifestations may not be discernible until the age of 60 or older.

Hereditary, heritable, and *inherited* are essentially synonymous terms. A case of hereditary disease occurring as the result of a new mutation is heritable, although not inherited unless one considers that the occurrence of the mutation in the germ cell of a parent and subsequent passage to the offspring constitutes inheritance.

Aside from new mutations there are other genetic disorders that are usually not inherited, namely many of the chromosomal aberrations. Some chromosomal aberrations, however, are inherited. For example, there is a 50–50 chance that an offspring of a Down's syndrome patient will also have Down's syndrome. The 15–21 translocation responsible for multiple cases of Down's syndrome in some families is hereditary.

Strictly speaking *familial* is not synonymous with *hereditary* or *genetic* since nongenetic factors can account for familial aggregation.

In the past *familial* was the term used for recessive disorders since they tend to occur in multiple sibs with normal parents and *hereditary* was the term used for dominant disorders. Obviously, a recessive disorder is as genuinely inherited as is a dominant one. Recessive inheritance is inheritance from both parents.

The Causes of Birth Defects

Several congenital malformations have a relatively simple genetic basis, for example, autosomal recessive acheiropody (absence of hands and feet), which has been observed only in Brazil (Fig. 5.5).

A few congenital malformations have a relatively simple environmental or extrinsic cause. Two examples are the malformations of the heart, eyes, and other organs due to rubella (German measles) occurring in the first 12 weeks of pregnancy, and phocomelia ("seal limbs") and other anomalies due to the consumption of thalidomide, a tranquilizer and sedative, during early pregnancy.

The majority of congenital malformations, for example, cleft palate, harelip, clubfoot, anencephaly, and congenital heart malformations, are probably the result of a collaboration of genetic and environmental factors, none of which is understood in any great detail.

FIG. 5.5. (a) *Acheiropody, a recessive trait that to date has been observed only in Brazil.* (b) *Pedigree. Courtesy of A. Freire-Maia.*

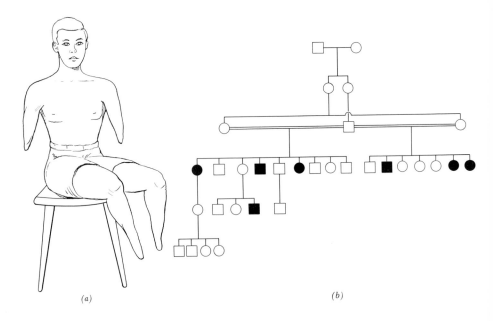

(a) *(b)*

Experiments with teratogenic agents in animals suggest that vulnerability to the effects of these agents is genetically determined. For example, cortisone will induce cleft palate in a high proportion of offspring in one strain of mice and in only a low proportion in another strain. The likelihood of producing a particular malformation with the administration of a teratogenic agent to a pregnant animal is often related to the frequency of that malformation as a "spontaneous" occurrence in the particular strain of animal. Nothing is known about genetic differences in susceptibility to chemical teratogens in man. In fact, few chemicals teratogenic to man have been demonstrated. This does not mean, of course, that many do not exist. Congenital malformations due to thalidomide were rather quickly recognized because of the dramatic nature of the malformation, phocomelia, which, furthermore, occurs very rarely as a "spontaneous" malformation. On the other hand, as much as a tenfold increase in the frequency of a more commonplace malformation such as cleft palate or harelip could be produced by a given chemical and not be recognized.

The importance of genetic factors in congenital malformations is underscored by the pronounced species differences in teratogenesis. Many agents, teratogenic in certain experimental animals, even in small dosage, give no evidence of being teratogenic in man. The "animal screening" of new drugs for possible teratogenic effects when administered to pregnant women is difficult because of these differences in species. Possibly studies in nonhuman primates will give more pertinent information on teratogenic effects of chemicals used in man.

Some specific human malformations have been observed to occur with greater frequency in some races than in others. For example, polydactyly is about 10 times more frequent in Negroes than in whites, and preauricular sinus may be about equally more frequent in Negroes. On the other hand, severe malformation of the nervous system, for example, anencephaly, is rarer in Negroes than in whites. The *total number* of congenital malformations—a substantial amount of data is available for Caucasians, Negroes, and Japanese—tends to be about the same in various races.

References

Benirschke, Kurt, "Accurate Recording of Twin Placentation: A Plea to the Obstetrician," *Obstet. Gynecol., 18* (1961), 334–47.

Bergsma, D., ed., *Conjoined Twins*. New York: National Foundation—March of Dimes, 1967.

Corner, G. W., "The Observed Embryology of Human Single-Ovum Twins and Other Multiple Births," *Am. J. Obstet. Gynecol., 70* (1955), 933–51.

Fishbein, Morris, ed., *Birth Defects*. Philadelphia: J. B. Lippincott Co., 1963.

Herschler, M. S., and N. S. Fechheimer, "The Role of Sex Chromosome Chimerism in Altering Sexual Development of Mammals," *Cytogenetics, 6* (1967), 161–67.

Markert, Clement, and Heinrich Ursprung, *Developmental Genetics*. Englewood Cliffs, N.J.: Prentice-Hall, Inc., 1969.

Potter, Edith L., "Twin Zygosity and Placental Form in Relation to the Outcome of Pregnancy," *Am. J. Obstet. Gynecol., 87* (1963), 566–77.

Genes in Families and in Population:
Mathematical Aspects

Hardy and Weinberg considered gene frequency the pertinent variable in population genetics. The alleles at a given locus are thought of as occurring in a "gene pool." Under conditions of random mating, as is assumed by the Hardy–Weinberg principle, two alleles meet in the diploid organism in frequencies that are the product of the individual gene frequencies. For example, in a two-allele system if p is the relative frequency (or proportion) of the dominant allele A, q the relative frequency (or proportion) of the recessive allele a, and $p + q = 1$, then in a randomly mating population the frequencies of the three genotypes are p^2 (for AA), $2pq$ (for Aa), and q^2 (for aa). (See Mettler and Gregg's *Population Genetics and Evolution* in this series.)

The Hardy–Weinberg principle, the cornerstone of population genetics, states that the relative proportions of genotypes with respect to one autosomal locus remain constant from one generation to another—for example, the three genotypes AA, Aa, and aa in a system with two alleles A and a. Like all generalizations in science, this one is based on certain simplifying assumptions and disregards certain factors (such as mutation and selection) that disturb the equilibrium. The Hardy–Weinberg equilibrium is the basis for a consideration of the influence of these factors on the relative proportions of the three genotypes in successive generations. The algebraic

TABLE 6.1.   *The Hardy–Weinberg Equilibrium*

(a) *Characteristics of the parental generation*

Allelic genes	Assigned frequency	Numerical example	Genotypes	Frequencies in population	Numerical example
A	p	0.90	AA	p^2	81
a	q	0.10	Aa	$2pq$	18
			aa	q^2	1
TOGETHER		1.00	ALL TYPES	1	100

(b) *Offspring from random matings*

Parents ♂ ♀	Frequency of mating type	Frequency of offspring*			Numerical example		
		AA	Aa	aa	AA	Aa	aa
$AA \times AA$	p^4	p^4	—	—	6,561	0	0
$AA \times Aa$ $Aa \times AA$	$4p^3q$	$2p^3q$	$2p^3q$	—	1,458	1,458	0
$AA \times aa$ $aa \times AA$	$2p^2q^2$	—	$2p^2q^2$	—	0	162	0
$Aa \times Aa$	$4p^2q^2$	p^2q^2	$2p^2q^2$	p^2q^2	81	162	81
$Aa \times aa$ $aa \times Aa$	$4pq^3$	—	$2pq^3$	$2pq^3$	0	18	18
$aa \times aa$	q^4	—	—	q^4	0	0	1
ALL TYPES	1	p^2	$2pq$	q^2	8,100	1,800	100

* The AA column adds up to $p^4 + 2p^3q + p^2q^2$, or $p^2(p^2 + 2pq + q^2)$, or p^2.
The Aa column adds up to $2p^3q + 4p^2q^2 + 2pq^3$, or $2pq(p^2 + 2pq + q^2)$, or $2pq$.
The aa column adds up to $p^2q^2 + 2pq^3 + q^4$, or $q^2 + 2pq + q^2$, or q^2.

example in Table 6.1 shows that genotype frequencies indeed remain constant from one generation to the next.

The analytic usefulness of the fundamental proportions of genotypes $(p^2 : 2pq : q^2)$ is illustrated by the fact that by knowing the frequency of the recessive phenotype (q^2) one can calculate the proportion of heterozygotes $(2pq)$, which is relatively large even if the recessive phenotype is rare. In the example of Table 6.1 the homozygotes aa are only 1 in 100, but the heterozygotes Aa are 18 in 100. Albinism, a recessive trait, occurs about 1 in 10,000 persons (q^2). The frequency of the recessive allele (q) is then $\sqrt{1/10,000}$, or $1/100$. The frequency of the dominant allele (p) is $1 - 1/100$, or $99/100$. The frequency of heterozygous carriers of albinism $(2pq)$ is $2 \times 1/100 \times 99/100$, or about 1 in 50.

When the persons heterozygous for a dominant gene cannot be distinguished from those homozygous, gene frequency must be calculated from the frequency of recessive homozygotes, as just illustrated for the case of albinism. However, when all three genotypes are separately identifiable, a counting method can be used to estimate gene frequency. For example, it was found among 1,279 persons that 363 were type M, 634 were MN and 282 were N. Consequently, out of the total of $1,279 \times 2$, or 2,558 genes, 726 (363×2) + 634 are M; the frequency of the M gene is 0.532 (1,360/2,558).

For X-linked traits the phenotype frequency in males is equal to the gene frequency. In a two-allele system the male must be one of only two phenotypes and genotypes. The trait must either be present or be absent and the frequencies are p and q, respectively. It does not matter, furthermore, whether the trait is recessive or dominant in the female; in either case the phenotype frequency in the male is identical to the gene frequency.

Rare X-linked dominant traits occur about twice as frequently in females as in males. Females affected by a dominant X-linked trait are of two types—homozygotes with a frequency p^2 and heterozygotes with a frequency $2pq$. The frequency of affected males is p. The ratio of affected females to affected males is, then, $(p^2 + 2pq)/p$. For a rare X-linked dominant trait (such as vitamin D-resistant rickets) p^2 is negligibly small and q is almost 1. Therefore, the F : M ratio becomes about $2p/p$ (affected females twice as frequent as affected males). With more frequent X-linked dominants, such is no longer the case. Thus, for the Xg^a blood group, about 64 percent of men are $Xg(a+)$ and about 88 percent of women are $Xg(a+)$, giving an F : M ratio of about 1.4 : 1.

Mating Frequencies

Under conditions of random mating, which is an assumption of the Hardy–Weinberg principle, the frequency of matings of different types is the product of the frequencies of the individuals making up the mating. This is illustrated in part b of Table 6.1 and in Table 6.2. If p^2 is the frequency of AA persons, then $p^2 \times p^2$, or p^4, is the frequency of $AA \times AA$ matings. Since $2pq$ is the frequency of Aa persons, the frequency of $AA \times Aa$ matings is $p^2 \times 2pq \times 2$, or $4p^3q$. Note that the product is doubled in this case because the $AA \times Aa$ mating may be either $AA\ \male \times Aa\ \female$ or $AA\ \female \times Aa\ \male$. When the genotypes of the parents differ the product is doubled to arrive at the mating frequency.

When the homozygotes and heterozygotes have indistinguishable

TABLE 6.2. *Frequency of Mating Types**

Mating type			Relative frequency	
Father A		Mother B	$A \times B$	Simplest expression
I	Dominant phenotype $p^2 + 2pq$	× Dominant phenotype $p^2 + 2pq$	$p^4 + 4p^3q + 4p^2q^2$	$p^2(p+2q)^2 = p^2(1+q)^2$†
II	Dominant phenotype $p^2 + 2pq$	× recessive phenotype q^2	$2p^2q^2 + 4pq^3$	$2pq^2(p+2q) = 2pq^2(1+q)$†
	Recessive phenotype q^2	× dominant phenotype $p^2 + 2pq$		
III	Recessive phenotype q^2	× recessive phenotype q^2	q^4	q^4

* See Table 6.1b for a different presentation and Table 6.3 for analysis of data.
† Since $p = 1 - q$, $p^2(p+2q)^2 = p^2(1+q)^2$, and $2pq^2(p+2q) = 2pq^2(1+q)$.

128

phenotypes, only three types of mating are identifiable. These are shown in Table 6.2.

Working Out the Mode of Inheritance of a Polymorphic Trait

The approach to determining the mode of transmission of a polymorphic trait is well illustrated by studies of the ability to taste phenylthiocarbamide (PTC) or the lack of this ability. About 40 years ago a chemist at the Dupont Laboratories discovered by chance that some people found PTC bitter to the taste whereas others experienced no bitter taste. The heritability of the trait and its precise mode of inheritance were determinable from family studies. Only two classes of individuals—tasters and nontasters—were identifiable. If nontasting was the recessive phenotype, then tasting must have two alternative genotypic bases, the homozygous and the heterozygous states, but these were not distinguishable.

The data used here are those concerning 800 families tested in Ohio by Laurence H. Snyder. Among the 1,600 parents, 1,139 were tasters and 461 were nontasters. On the assumption that nontasting is recessive, the frequency of that phenotype (q^2) is 461/1,600, or 28.8 percent, and the frequency of the nontaster allele (q) is $\sqrt{0.288}$, or 0.537. The frequency of the taster allele (p) is $1 - 0.537$, or 0.463. (When all 3,643 parents and offspring in the 800 families were considered together, the frequency of nontasters was 29.8 percent. It is theoretically preferable, however, to use only the parents in the calculation of gene frequency because for the most part these represent a collection of unrelated persons.) Note that although the nontaster phenotype is the rarer of the two the nontaster gene is the more frequent one if the recessive hypothesis is true.

The next step is to determine whether the proportions of taster and nontaster offspring from matings of the three types (shown in Table 6.2) correspond with Mendelian expectations. First it should be determined whether the matings are random as far as taste-test is concerned. Do the proportions of the three types agree with the proportions predicted by Table 6.2? This analysis is shown in Table 6.3 from which it is concluded that mating is nonassortative in regard to PTC tasting.

Having determined that matings are random as evidenced by the fact that they follow a binomial distribution, we next determine whether the proportions of taster and nontaster offspring in families of the three parental mating types correspond with Mendelian expectation. What is that expectation? Obviously all children of recessive × recessive parents should be recessive. The phenotypic proportions of offspring of the other two types of matings are not so easily predicted. Some of the tasters are homozygous, some are heterozygous.

TABLE 6.3. *Parental Mating Types in 800 Families Tested for PTC Tasting*

| | Expected* frequency of mating | | Observed | |
	General expression	Frequency	Number	Frequency
Phenotype of mating				
Taster × taster	$p^2(1+q)^2$	0.506	425	0.531
Taster × nontaster	$2pq^2(1+q)$	0.410	289	0.361
Nontaster × nontaster	q^4	0.083	86	0.108

SOURCE: Laurence H. Snyder, *Ohio J. Science, 32* (1932) , 436.
* Assuming that the nontaster has the recessive phenotype and that the frequency of the taster and nontaster alleles (p and q) are 0.463 and 0.537, respectively.

Shown in Table 6.4 are the algebraic expressions for the proportions of children expected with each phenotype according to the mating type of the parents. Among the offspring of dominant × recessive matings, the proportion with the recessive phenotype is

$$\frac{2pq^3}{2p^2q^2 + 4pq^3}$$

Dividing both the numerator and denominator by $2pq^2$, we get

$$\frac{q}{p + 2q}$$

which we can express entirely in terms of q, because $p = 1 - q$; we then obtain

$$\frac{q}{1 - q + 2q} = \frac{q}{1 + q}$$

Similarly, the proportion of offspring with the recessive phenotype produced by dominant × dominant matings is

$$\frac{p^2q^2}{p^4 + 4p^3q + 4p^2q^2}$$

Dividing both numerator and denominator by p^2, we get

$$\frac{q^2}{p^2 + 2pq + 4q^2} = \left(\frac{q}{p + 2q}\right)^2$$

TABLE 6.4. *Phenotype Frequencies in Children*

	Mating types ♂ ♀	Parental genotype frequencies	Offspring			Proportion of recessive offspring
			AA	Aa	aa	
I	D × D					
	$AA \times AA$	p^4	p^4	—	—	$\left(\dfrac{q}{1+q}\right)^2$
	$AA \times Aa$ ⎫ $Aa \times AA$ ⎬	$4p^3q$	$2p^3q$	$2p^3q$	—	
	$Aa \times Aa$	$4p^2q^2$	p^2q^2	$2p^2q^2$	p^2q^2	
II	D × R					
	$AA \times aa$ ⎫ $aa \times AA$ ⎬	$2p^2q^2$	—	$2p^2q^2$	—	$\dfrac{q}{1+q}$
	$Aa \times aa$ ⎫ $aa \times Aa$ ⎬	$4pq^3$	—	$2pq^3$	$2pq^3$	
III	R × R					
	$aa \times aa$	q^4	—	—	q^4	100%

131

TABLE 6.5. *Number of Nontaster Offspring, According to Parental Mating Type, in 800 Families*

Parental mating type	No. of children	Expected* proportion of nontaster offspring		No. of nontaster offspring	
		Formula	Specific value	Ex-pected	Ob-served
Taster × taster	1,059	$\left(\dfrac{q}{1+q}\right)^2$	0.122	129.3	130
Taster × nontaster	761	$\dfrac{q}{1+q}$	0.349	265.9	278
Nontaster × nontaster	223	100%		223.0	218

SOURCE: Laurence H. Snyder, *Ohio J. Science, 32* (1932) , 436.
* Assuming that the nontaster has the recessive phenotype and that the frequency of the taster and nontaster alleles (*p* and *q*) is 0.465 and 0.535, respectively.

If, as before, we substitute $1 - q$ for *p*, this becomes

$$\left(\frac{q}{1-q+2q}\right)^2 = \left(\frac{q}{1+q}\right)^2$$

$q/(1+q)$ and $[q/(1+q]^2$ are called *Snyder's ratios,* having been devised by Laurence H. Snyder, whose data on PTC tasting are used here as an example.

In Table 6.5, data on the offspring of the three types of parental matings are shown and the proportions of children of recessive phenotype are compared with those expected according to Snyder's ratios. Agreement of "observed" with "expected" is satisfactory.

Note that five tasters occurred among the offspring of nontaster × nontaster parents, whereas no taster offspring are predicted by the recessive hypothesis. These five exceptions may reflect inaccuracy in phenotyping or perhaps illegitimacy (mistaken paternity) .

The reader should perform the same calculations, assuming that the taster phenotype is recessive. These calculations show poor fit of observed with expected.

Other methods for testing the segregation are given in texts on genetical statistics. Also see Race and Sanger (1968) for illustration of the use of Fisher's method, in the analysis of data on the P blood groups.

Bias of Ascertainment

A skillful college genetics teacher announced to his all-male class of 129 students that they would determine the normal human sex ratio by adding the numbers of males and females in the sibships from which each of them came. The data were recorded as shown in Table 6.6.

The professor used this example to illustrate bias of ascertainment: when sibships are ascertained because of the occurrence in them of one person with a given trait (in this example, maleness was the trait) the proportion of persons affected by said trait in the families cannot be expected to agree with the true proportion if the proband is included in the enumeration. Inclusion of the proband loads the results in favor of the trait. If all human sibships had 100 individuals or even 10, the effect would be much less pronounced. But obviously all *ascertainable* one-child sibships will have 100 percent trait-bearers, two-child sibships will have 67 percent trait bearers (when the fundamental probability is 50 percent), three-child sibships will have 57 percent trait bearers, and so on.

TABLE 6.6. *A Class Experiment**

	Male	Female
Abrams	1	1
Adams	1	0
Allen	3	2
Anderson	2	1
—	—	—
—	—	—
—	—	—
Young	2	0
Ziegler	1	2
TOTAL	228	95
GRAND TOTAL		323

* What, the professor asked his class, is the sex ratio in this sample of families? The first reply was 228/95, a preposterously high sex ratio.

The true sex ratio is obtained by removing the probands from the calculation. The ratio then becomes 99/95, or 1.04.

Testing the Recessive Hypothesis

In testing the segregation of a rare autosomal dominant trait, if families are ascertained through an affected parent, it is a simple matter to determine whether the proportion of affected children agrees with the expected 50 percent (only one parent being affected). But usually families are ascertained in a variety of ways, perhaps most often through an affected child. In sibships so detected the proportion of affected sibs will, of course, exceed 50 percent if the proband is counted in. So bias of ascertainment must be kept in mind. However, it is in the study of rare autosomal recessive traits that the bias of ascertainment represents the most significant problem. This is because one usually can recognize those families that have both parents heterozygous and therefore have a 25 percent chance of having affected children *only* when at least one child is affected. Those parental couples who, although both heterozygous for the recessive gene, by good fortune had no affected children will not be represented in the collection of families.

Consider a group of 16 families with 2 children in each and with both parents heterozygous for a rare recessive gene (Fig. 6.1*a*). In these 16 families the expectation is that 4 of the 16 first-born children will be affected. Likewise, chances are that 4 of the 16 second-born will be affected, but since this is an event independent of the first, in only one of the 16 families ($\frac{1}{4} \times \frac{1}{4} = \frac{1}{16}$) will *both* children be affected. These expectations are schematized in a second way in Fig. 6.2.

Since the mode of ascertainment is through affected children, only 7 of the 16 families can be recognized (Fig. 6.1*b*). The proportion of affected children in these families is found to be not $\frac{1}{4}$ but $\frac{8}{14}$ (57 percent) (Fig. 6.1*c*). The general expression for unascertained sibships is $(\frac{3}{4})^s$, where s is sibship size and $\frac{3}{4}$ is the probability of the dominant phenotype. In Fig. 6.3, segregation in three-child families (with both parents heterozygous) is illustrated.

It will be noted that the unascertained sibships are cut off the end of the binomial expansion $(d + r)^s$ where d is the probability of the dominant phenotype (in this instance $\frac{3}{4}$), r the probability of the recessive phenotype ($\frac{1}{4}$), and s the sibship size. Figure 6.4 gives the formulas and the proportions of families with various numbers of affected children, for families of 2, 3, and 4 sibs. For five-sib families the formula is as follows:

$$\left(\tfrac{3}{4}\right)^5 + 5\left(\tfrac{3}{4}\right)^4\left(\tfrac{1}{4}\right) + 10\left(\tfrac{3}{4}\right)^3\left(\tfrac{1}{4}\right)^2 + 10\left(\tfrac{3}{4}\right)^2\left(\tfrac{1}{4}\right)^3 + 5\left(\tfrac{3}{4}\right)\left(\tfrac{1}{4}\right)^4 + \left(\tfrac{1}{4}\right)^5$$

and so on.

The coefficients in the binomial expansion are easily recalled from Pascal's triangle:

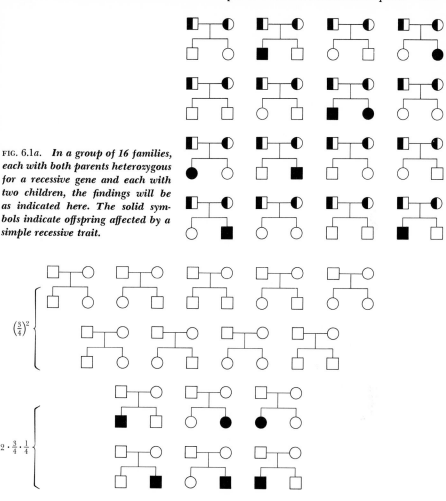

FIG. 6.1*a*. **In a group of 16 families, each with both parents heterozygous for a recessive gene and each with two children, the findings will be as indicated here. The solid symbols indicate offspring affected by a simple recessive trait.**

FIG. 6.1*b*. **Only 7 of the 16 families will be ascertained when the families are found because of at least one affected child, as is usually the case.**

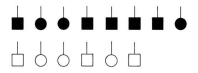

FIG. 6.1*c*. **The proportion of children affected in the ascertained families is not 1 in 4 but rather 8 in 14 (57 percent).**

```
                        1
                  1           1
              1       2           1
          1       3       3           1
      1       4       6       4           1
  1       5      10      10       5           1
1     6      15      20       15      6          1
                      etc.
```

Or they can be derived from the formula

$$\frac{s!}{a!\,b!}$$

where s is the total number of sibs and a and b are the number of sibs of each of the two types. Because the first term of the binomial expansion is cut off, detection of families through affected sibs is called truncate ascertainment.

If every case of a rare autosomal recessive trait were ascertained in a population (or at least a random collection of affected families), then the distribution of families according to proportions with the several different numbers of children affected would be as indicated here. Using the principles just outlined, several early students of human genetics devised methods for testing the observed proportions of affected sibs with those expected if the trait under study is autosomal recessive. The proportion of affected children differs with families of different size. One counts up the total number of children and the number affected in sibships of the various sizes; these figures are the "observed." One then multiplies the total number of sibships of a particular size by the proportion expected to be affected with that size of sibship (sz); this is the "expected." "Observed" is then compared with "expected." Table 6.7 provides the values used in calculating the expected. In addition, the variance for each value is given. Since the outcomes in the families are independent, the variance of the sum is the sum of the variances.

First child

	Normal	Normal	Normal	Affected
Normal				One affected
Normal				One affected
Normal				One affected
Affected	One affected	One affected	One affected	Both affected

(Second child)

FIG. 6.2. *Expected distribution of a recessive trait in two-child families, when both parents are heterozygous.*

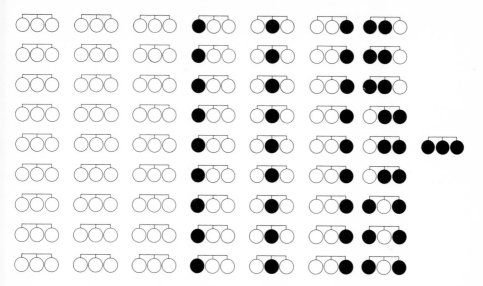

FIG. 6.3. *Segregation of a simple recessive trait (solid symbols) in families of three children with both parents heterozygous. Here and in Fig. 6.7 the symbols may, of course, represent either males or females. Redrawn from C. C. Li,* **Human Genetics** *(New York: McGraw-Hill, 1961).*

FIG. 6.4. *Various proportions of sibs affected by an autosomal recessive trait in families of various sizes with both parents heterozygous.*

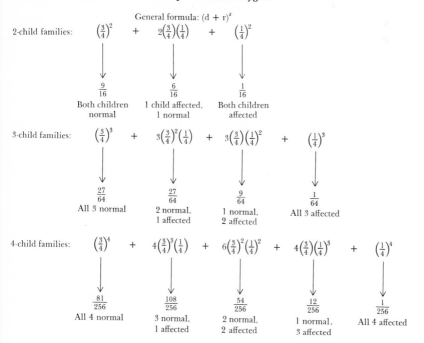

TABLE 6.7. *Predicted Values for Testing Recessive Hypothesis Under Conditions of Complete Ascertainment or Random Selection of Affected Families*

Sibship size (s)	Proportion of sibships that cannot be ascertained $[x = (3/4)^s]$		Proportion of sibships that can be ascertained $[y = 1 - (3/4)^s]$	Proportion of affected in ascertained sibships (z)	Average number affected (sz)	Variance of number affected
1		3/4	1/4	1	1	0.000
2	$(3/4)^2 =$	9/16	7/16	8/14	1.143	0.122
		(0.5625)	(0.4375)	(0.5714)		
3	$(3/4)^3 =$	27/64	37/64	48/111	1.297	0.263
		(0.4219)	(0.5781)	(0.4324)		
4	$(3/4)^4 =$	81/256	175/256	256/700	1.463	0.420
		(0.3164)	(0.6836)	(0.3657)		
5	$(3/4)^5$	(0.2373)	(0.7627)	(0.3278)	1.639	0.502
6	$(3/4)^6$	(0.1780)	(0.8220)	(0.3041)	1.825	0.776
7	$(3/4)^7$	(0.1335)	(0.8665)	(0.2885)	2.020	0.970
8	$(3/4)^8$	(0.1001)	(0.8999)	(0.2778)	2.223	1.172
9	$(3/4)^9$	(0.0751)	(0.9249)	(0.2703)	2.433	1.380
10	$(3/4)^{10}$	(0.0563)	(0.9437)	(0.2649)	2.649	1.592
11	$(3/4)^{11}$	(0.0422)	(0.9578)	(0.2610)	2.871	1.805
12	$(3/4)^{12}$	(0.0317)	(0.9683)	(0.2582)	3.098	2.020
13	$(3/4)^{13}$	(0.0238)	(0.9762)	(0.2561)	3.329	2.234
14	$(3/4)^{14}$	(0.0178)	(0.9822)	(0.2545)	3.563	2.446

Application of the above method for testing the recessive hypothesis (it is called the *a priori method,* or *method of Apert*) can be illustrated with the data on a rare malformation syndrome, the Ellis–van Creveld (EvC) syndrome, or six-fingered dwarfism (Fig. 6.5). This condition is unusually frequent among the Old Order Amish of Lancaster County, Pennsylvania. An attempt was made to ascertain all cases in the community. Affected persons were found in 33 separate sibships, as indicated in Fig. 6.6. One of the 33 sibships (No. 22 in Fig. 6.6) had an affected father; the mother was presumably heterozygous. This sibship is, of course, excluded from the analysis.

In the 32 *ascertainable* sibships there is a total of 200 children. In each family both parents are phenotypically normal but by the recessive hypothesis must be heterozygous. The proportion of affected children from two heterozygous parents is one-fourth. How many of the 200 children will then be affected? Fifty? No, appreciably more than 50 (25 percent) will be affected, for as is explained above, as-

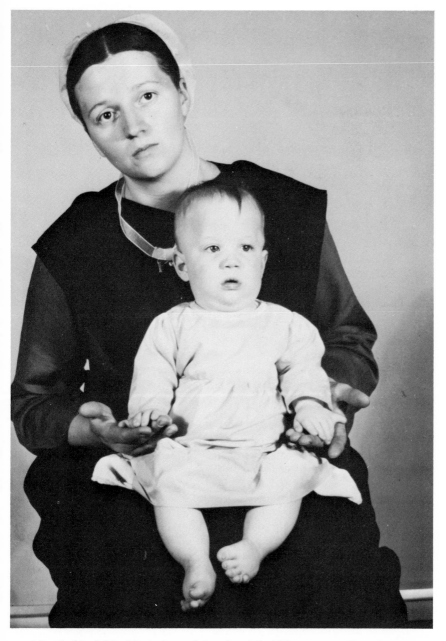

FIG. 6.5.   *Amish child with six-fingered dwarfism (the Ellis–van Creveld syndrome).*

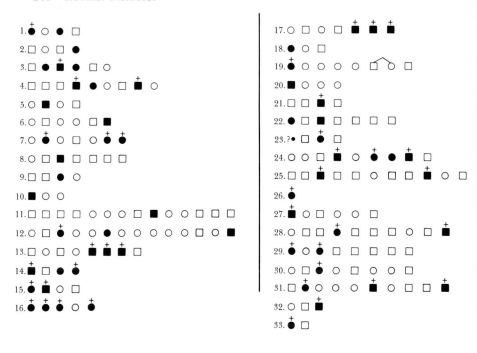

FIG. 6.6. *Amish sibships with one or more cases of the Ellis–van Creveld syndrome. All but No. 22 had normal parents. The father of sibship 22 was affected.*

certainment is truncate. All those families of heterozygous × heterozygous parents with no affected children were not ascertained.

When the data on the other 32 sibships (with both parents normal but presumably heterozygous for the EvC gene) are analyzed, comparing observed with expected as indicated by Table 6.7, the results shown in Table 6.8 are obtained. The standard deviation (S.D.) is the square root of the variance. Since observed deviates from expected by only a fraction of one standard deviation, the deviation is not significant and the recessive hypothesis is supported. Because of the large mean family size of over 6 children (and some families are not yet complete), the observed proportion of affected children in the ascertained sibships (59/200, or 29.5 percent) does not deviate as much from 25 percent as would be the case in a collection of American families of the more common smaller size.

TABLE 6.8.   *The Ellis–van Creveld Syndrome*

Sibship size	No. of sibships	No. of sibs	Number affected		Variance
			Observed	Expected	
1	1	1	1	1.0000	0.0
2	1	2	1	1.1428	0.12245
3	4	12	4	5.1892	1.05188
4	8	32	12	11.7024	3.36040
5	1	5	4	1.6389	0.50178
6	3	18	5	4.9167	2.32770
7	3	21	7	6.0588	2.91072
8	4	32	7	8.8900	4.68960
9	2	18	7	4.8656	2.76040
10	2	20	5	5.2980	3.18340
11	1	11	2	2.8710	1.80530
14	2	28	4	7.1260	4.89280
TOTAL	32	200	59	61.6994	27.60643
					S.D. = 5.254

Incomplete Ascertainment

Note that the above method for correcting for "bias of ascertainment" is appropriate when ascertainment is complete or when a random sample of affected sibships is ascertained. In the study of six-fingered dwarfism in the Amish all or almost all cases were thought to have been ascertained and the close agreement of observed with expected corroborates not only the recessive mode of inheritance but also (admittedly with a certain circularity in the logic) the completeness of ascertainment.

Often, however, ascertainment of affected families is not only incomplete but also is nonrandom. The probability that a given family enters the sample often bears some relationship to the number of sibs affected. For example, among three-child sibships, one with all three affected by a given rare recessively inherited disease may stand a much greater chance than one with only a single affected child of coming to the attention of facilities the records of which are screened for cases of that disease, for example, hospitals, crippled children's agencies, and practicing specialists.

If a sample of families were collected by random selection of affected *individuals* and the probability of any one case being ascertained were small, the probability of a given family finding its way into a series of cases would be directly proportional to the number of

TABLE 6.9. *Predicted Number of Affected Sibs, When Probability of Ascertainment of Family Is Directly Related to the Number of Affected Sibs*

Size sibship		Number of sibships	Number of children	Number affected	Predicted affected	
					Proportion of sibs	Average number per sibship
2	(1) 2 $(3/4)(1/4)$	6	12	6		
	(2) 1 $(1/4)^2$	2	4	4		
			16	10	0.625	1.25
3	(1) 3 $(3/4)^2(1/4)$	27	81	27		
	(2) 3 $(3/4)(1/4)^2$	18	54	36		
	(3) 1 $(1/4)^3$	3	9	9		
			144	72	0.500	1.50
4	(1) 4 $(3/4)^3(1/4)$	108	432	108		
	(2) 6 $(3/4)^2(1/4)^2$	108	432	216		
	(3) 4 $(3/4)(1/4)^3$	36	144	108		
	(4) 1 $(1/4)^4$	4	16	16		
			1,024	448	0.438	1.75
5	(1) 5 $(3/4)^4(1/4)$	405	2,025	405		
	(2) 10 $(3/4)^3(1/4)^2$	540	2,700	1,080		
	(3) 10 $(3/4)^2(1/4)^3$	270	1,350	810		
	(4) 5 $(3/4)(1/4)^4$	60	300	240		
	(5) 1 $(1/4)^5$	5	25	25		
			6,400	2,560	0.400	2.00
	etc.	etc.	etc.	etc.	etc.	etc.

affected individuals in the sibship. In predicting the "expected" proportions under this condition of so-called single ascertainment, one goes through the same procedure as that for deriving Table 6.7, with the additional step that the terms in the binomial expansion are multiplied by the number of affected sibs, as shown in Table 6.9. In Table 6.10 the values are given for sibships up to 13 in size.

It can be shown that removal of one affected sib from each ascertained family corrects properly for the bias of ascertainment involving random selection of affected persons. This is shown schematically in Fig. 6.7. In *a,* the sibships are represented in proportions directly related to the number of affected sibs. In *b,* one affected sib has been removed from each sibship and it can be seen that exactly one-fourth of the sibs are affected. The general formula for the estimated segregation ratio in what is called the *simple sib method* is

TABLE 6.10. *Predicted Values for Testing Recessive Hypothesis Under Conditions of Incomplete Single Ascertainment, or Random Selection of Affected Individuals*

Sibship size (s)	Proportion of affected in ascertained sibships (z)	Average number affected (Sz)
1	1.000	1.00
2	0.625	1.25
3	0.500	1.50
4	0.438	1.75
5	0.400	2.00
6	0.375	2.25
7	0.357	2.50
8	0.344	2.75
9	0.333	3.00
10	0.325	3.25
11	0.318	3.50
12	0.313	3.75
13	0.308	4.00

$$\hat{q} = \frac{R - N}{T - N}$$

where R is the total number of persons affected by the recessive condition (72, in the example of Fig. 6.7); N is the number of ascertained sibships (in the example, 48); and T is the total number of sibs (in the example, 144). Thus, in the example, the estimated segregation ratio is:

$$\hat{q} = \frac{72 - 48}{144 - 48} = \frac{24}{96} = \frac{1}{4}$$

In the approaches shown in Tables 6.7 and 6.10, the expected values are adjusted, according to the appropriate type of ascertainment, to permit a comparison with "observed." In the approach just given, the observational data are corrected to test the fit with the quarter ratio. A similar approach (of adjusting the data to arrive at a segregation ratio) can be used with the assumption of complete ascertainment (or random ascertainment of families).

In practice, ascertainment is neither complete (for which the a priori, or Apert, correction is appropriate) nor related as a simple integer to the number of affected sibs (for which the simple sib method is

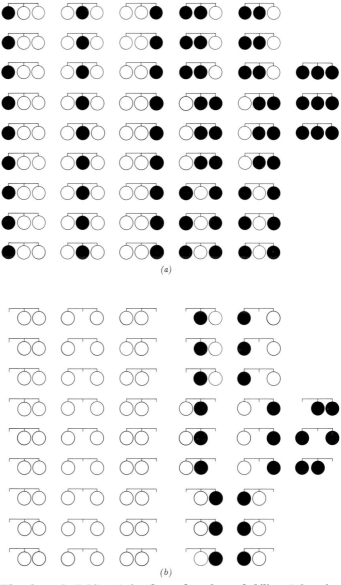

FIG. 6.7. (a) *The observed sibships of size three when the probability of detecting sibships with two and three members affected by a recessive trait is, respectively, two and three times the probability of detecting a sibship with one affected member. This is termed single ascertainment.* (b) *The method of correction is to delete one affected member from each sibship. Thus, 24 affected among a total of 96 children, or 1 in 4, is found. Redrawn from* C. C. Li, Human Genetics (*New York: McGraw-Hill, 1961*).

appropriate). Analysis of the data may show results intermediate between those of the two methods, suggesting that ascertainment was of an intermediate type partaking of features of both polar types.

Quantitative Aspects of Consanguinity

Mating is nonrandom, or assortative, if the probability of a particular type of mating is not dictated solely by the relative frequencies of the three genotypes in the population. Consanguineous marriages are nonrandom because the genotypes of closely related persons have an increased chance of being similar rather than representative of the general population.

The coefficient of consanguinity, or coefficient of inbreeding (F), is the probability that any single locus is homozygous by descent from a common ancestor. "By descent from a common ancestor" (*autozygous* is a useful word for this type of homozygosity) is an important part of the definition because homozygosity on other bases is not included in the estimate.

Calculation of the coefficient of consanguinity for the offspring of a first-cousin marriage can be illustrated with the diagram shown in Fig. 6.8a. Alleles W, X, Y and Z are present in the great-grandparents of No. 1, the person whose inbreeding is being calculated. The probability that No. 1 received his great-grandfather's gene W from his father is $\frac{1}{2} \times \frac{1}{2} \times \frac{1}{2}$, or $\frac{1}{8}$. The probability that he got it from his mother is also $\frac{1}{8}$. Therefore, the probability that he is autozygous for gene W is $\frac{1}{8} \times \frac{1}{8}$, or $\frac{1}{64}$. Similarly, the probability that he is autozygous for X is $\frac{1}{64}$; that he is autozygous for Y is $\frac{1}{64}$; and that he is

FIG. 6.8. *Calculation of coefficient of consanguinity of an offspring of a first-cousin marriage.* (a) *Diagram illustrating the principle.* (b) *Diagram illustrating Wright's method of path coefficients.*

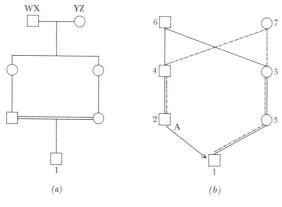

(a)   (b)

autozygous for Z is $\frac{1}{64}$. The probability that he is autozygous for *one* of the four genes is the sum of the four individual probabilities, or 1 in 16.

In actual calculation, especially when the relationship of the parents is complicated (for example, they may be not only first cousins, but also second cousins through three connections and third cousins through several connections), Sewall Wright's method of path coefficients has been found useful. Its use can be illustrated with the diagram shown in Fig. 6.8*b*. It is a simpler method than the method previously described because it identifies all common ancestors of the two parents and totals, for each common ancestor, the steps in the path connecting one parent to the inbred offspring through the common ancestor.

Assuming that the proband (No. 1) in Fig. 6.8*b* received allele A from the father (No. 2), what is the chance that he received the same allele from the mother (No. 3)? The chance that the father got the gene from the grandfather (No. 4) is $\frac{1}{2}$. Assuming each of these individuals possessed allele A, the chance that No. 6 gave it to No. 5 is $\frac{1}{2}$; the chance that No. 5 gave it to No. 3 is $\frac{1}{2}$ and the chance that No. 3 gave it to No. 1 is $\frac{1}{2}$. The value $(\frac{1}{2})^5$ is the probability that locus A in person No. 1 is autozygous by descent from the great-grandfather. But there is also a probability of $(\frac{1}{2})^5$ that the locus is autozygous by descent from the great-grandmother (as diagrammed by the dashed line in Fig. 6.9*b*). The joint probability is the sum $(\frac{1}{2})^5 + (\frac{1}{2})^5$, or $(\frac{1}{2})^4$, or $\frac{1}{16}$. (Figure 6.9 shows the calculation of coefficient of inbreeding of the offspring from certain other types of matings and Table 6.11 lists values for F.) The general expression for the coefficient of consanguinity is $F = \Sigma[(\frac{1}{2})^n]$, where n is the number of ancestors in the path and the values for all paths are summed.

TABLE 6.11. *Coefficient of Consanguinity* (**F**) *for Various Parental Relationships*

Parental relationship	F	
Father–daughter	1/4	= 0.25000
Offspring of identical twins	1/8	= 0.12500
Uncle–niece	1/8	= 0.12500
Double first cousins	1/8	= 0.12500
First cousins	1/16	= 0.06250
Half first cousins	1/32	= 0.03125
First cousins once removed	1/32	= 0.03125
Second cousins	1/64	= 0.01563
Second cousins once removed	1/128	= 0.00782
Third cousins	1/256	= 0.00391

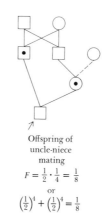

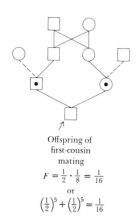

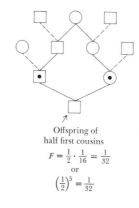

Offspring of
father-daughter
mating

$$F = \frac{1}{2} \cdot \frac{1}{2} = \frac{1}{4}$$

Offspring of
uncle-niece
mating

$$F = \frac{1}{2} \cdot \frac{1}{4} = \frac{1}{8}$$

or

$$\left(\frac{1}{2}\right)^4 + \left(\frac{1}{2}\right)^4 = \frac{1}{8}$$

Offspring of
first-cousin
mating

$$F = \frac{1}{2} \cdot \frac{1}{8} = \frac{1}{16}$$

or

$$\left(\frac{1}{2}\right)^5 + \left(\frac{1}{2}\right)^5 = \frac{1}{16}$$

Offspring of
half first cousins

$$F = \frac{1}{2} \cdot \frac{1}{16} = \frac{1}{32}$$

or

$$\left(\frac{1}{2}\right)^5 = \frac{1}{32}$$

FIG. 6.9. *The coefficient of inbreeding* (**F**) *of the offspring of certain consanguineous matings. The parents are indicated by the dot in the middle of the symbol.*

If the common ancestor is himself inbred, the path coefficient is multiplied by $(1 + F_A)$ where F_A is the ancestor's coefficient of consanguinity.

The coefficient of relationship (symbolized r) is the probability that two persons have in common a gene which came from the same ancestor. Stated differently, in two related persons, it is the proportion of all genes that are identical by descent. Estimation of the coefficient is made by the method of path coefficients, as for the coefficient of inbreeding. It can, perhaps, be seen that the coefficient of relationship for first cousins (see Fig. 6.10) is $(\frac{1}{2})^4 + (\frac{1}{2})^4$, or $\frac{1}{8}$. On the average, one-eighth of the genes of first cousins are identical, having been de-

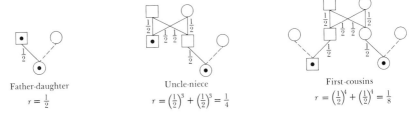

Father-daughter
$$r = \tfrac{1}{2}$$

Uncle-niece
$$r = \left(\tfrac{1}{2}\right)^3 + \left(\tfrac{1}{2}\right)^3 = \tfrac{1}{4}$$

First-cousins
$$r = \left(\tfrac{1}{2}\right)^4 + \left(\tfrac{1}{2}\right)^4 = \tfrac{1}{8}$$

FIG. 6.10. *The coefficient of relationship* (r) *of some other close relatives.*

rived from the same ancestor. The coefficient of consanguinity is always one-half the coefficient of relationship of the parents.

The frequency of consanguinity in the parents of cases of autosomal recessive disorders is inversely related to the gene frequency. The mathematical formulation for this relationship is as follows:

$$k = \frac{c\,(1 + 15q)}{16q + c\,(1 - q)}$$

where k tells us how often it can be expected that sibships containing individuals homozygous for a recessive allele have parents who are first cousins; c is the frequency of first-cousin marriages in the general population; and q is the frequency of the recessive allele. Because c and q are usually small, the approximate formula becomes simply $k = c/16q$.

In most of the United States today the frequency of first-cousin marriages is less than 1 in 1,000 (<0.1 percent). With a recessive gene whose frequency is 0.01 (the frequency of homozygous affected being 1 in 10,000), the frequency of first-cousin parental matings is about 0.7 percent. With a recessive gene whose frequency is 0.001, giving a frequency of the homozygous recessive of 1 in 1,000,000, the frequency of first-cousin parental matings is about 6 percent.

The relationship between gene frequency and the frequency of parental consanguinity is well illustrated by the findings in Tay-Sachs disease, a degenerative central nervous system disease that leads to early death. Inherited as an autosomal recessive, the disorder is rather frequent in Jewish persons, especially those whose ancestors lived in northeastern Poland and southern Lithuania. The gene frequency in the Jewish population of New York City appears to be between 0.011 and 0.016, with about 1 in every 30 persons being a carrier. Among the parents of Jewish children with Tay–Sachs disease there is little or no increase in the rate of consanguinity when the parents had Polish–Lithuanian ancestry. What is probably the same disease occurs much more rarely in non-Jewish persons, but in these persons there is an appreciable increase in the frequency of consanguinity of parents of affected children.

Knowing the frequency of parental consanguinity in a series of cases of an autosomal recessive trait, and knowing the frequency of consanguinity in the general population, one can make a crude estimate of gene frequency by rearranging the formula given above in this manner:

$$q = \frac{c}{16k - 15c}$$

In Japan a calculation of the frequency of the gene that in the homozygous state causes absence of catalase in the blood was based on an estimate of 0.06 for c and of 0.59 for k. Thus, the gene frequency may be of the order of 0.005.

The Probability That Twins Are Monozygotic

The probability of monozygosity may be increased, and thereby also the efficiency of the similarity method (of Sheila Maynard-Smith and Lionel S. Penrose), if the genotype of the parents with respect to objective marker traits is taken into account. (See Fig. 6.11.) The following question is posed: Given twin A as found and given the other

FIG. 6.11. *Are these twins monozygotic? See text. Redrawn from R. R. Race and R. Sanger,* **Blood Groups in Man, 4th ed. Oxford: Blackwell Scientific Publications, 1962.**

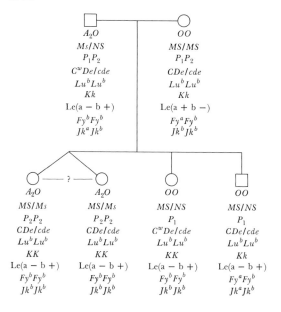

TABLE 6.12. *Probability of Monozygosity*

Chance of like-sex dizygotic twins	0.35	Chance of monozygotic twins	0.30
Chance of dizygotic twins in this family having same		Chance of monozygotic twins in this family having same	
ABO groups	0.50	ABO groups	1.00
MNSs groups	0.50	MNSs groups	1.00
P groups	0.25	P groups	1.00
Rh groups	0.25	Rh groups	1.00
Kell groups	0.25	Kell groups	1.00
Duffy groups	0.50	Duffy groups	1.00
Kidd groups	0.50	Kidd groups	1.00
Combined chance = product of separate chances	0.00034		0.30

relatives as found, what is the probability that the twin B should be found identical in all the phenotypes tested if in fact they are dizygotic?

In England, where the family diagrammed in Fig. 6.11 lived, about 35 percent of twins are dizygotic and of like sex. The chance (prior probability) that twin B would be the same sex as A is then 0.35. The mating $A_2O \times OO$ will produce 50 percent A_2 and 50 percent O children. The chance (conditional probability) of the second being A_2 like the first, even though they are dizygotic, is 0.50. The mating $Kk \times Kk$ will produce 25 percent KK offspring. The chance that the second twin will be KK, even if they are dizygotic, is then 0.25. The other conditional probabilities are worked out in Table 6.12. The combined chance (joint probability) is the product of the separate chances: for dizygosity 0.00034, and for monozygosity 0.30. The relative probability (also called the posterior probability) of monozygosity is

$$\frac{0.30}{0.30 + 0.00034} = 0.9989$$

and the relative probability of dizygosity is

$$\frac{0.00034}{0.30 + 0.00034} = 0.0011$$

When the genotypes of the parents are not known, the probability of monozygosity can be estimated by a comparable method based on gene frequencies in the population from which the twins are derived.

A comparable method is used to evaluate the probabilities in cases of disputed paternity.

Bayes' Principle in Genetic Prediction

The method applied above to estimating probability of mono-zygosity is that devised by the English clergyman-scientist Bayes and published in 1763. It considers two alternative possibilities: monozy-gosity and dizygosity. For each possibility, it considers the prior proba-bility that randomly selected like-sex twins will be of that zygosity. Then it considers the conditional probabilty that one would find the particular blood groups, given each zygosity. A joint probability is obtained by multiplying the prior probability and all of the condi-tional probabilities. The posterior, or relative, probability of each of the two alternative possibilities is obtained by dividing the joint probability by the sum of the two joint probabilities.

The same principle is used in estimating the chance that a particular woman is heterozygous for an X-linked recessive gene, such as that for muscular dystrophy. This application of the Bayes principle can be illustrated by means of the pedigree shown in Figure 6.12*a*. The con-sultand (person seeking advice) is marked *C*. The grandmother of the consultand has two affected sons and is therefore virtually certain to be a carrier. Thus the mother of the consultand has a prior probability of ½ of being a carrier. The fact that the mother had two normal sons reduces the likelihood that she is indeed a carrier, however. The con-sultand's question can be answered by the Bayesian calculations given in Table 6.13. The conditional probability is the probability that two sons would be normal under each of the two stated conditions. If the mother is heterozygous this value is ½ × ½; if the mother is not het-erozygous it is, of course, 1.

FIG. 6.12. *Situations for use of Bayes' theorem.*

■ Boy with X-linked muscular dystrophy

⊙ Female heterozygous for X-linked muscular dystrophy gene

(?) Female of uncertain status as to muscular dystrophy gene

Ⓒ The consultand

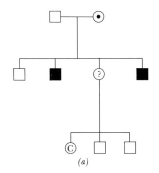

(a)

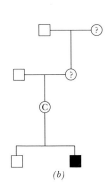

(b)

TABLE 6.13. *Bayesian Calculations Relevant to Fig. 6.12a*

	Consultand's mother heterozygous	Consultand's mother not heterozygous
Prior probability	$\frac{1}{2}$	$\frac{1}{2}$
Conditional probability	$(\frac{1}{2})^2 = \frac{1}{4}$	1
Joint probability	$\frac{1}{8}$	$\frac{1}{2}$
Posterior probability	$\dfrac{\frac{1}{8}}{\frac{1}{8} + \frac{1}{2}} = \frac{1}{5}$	$\dfrac{\frac{1}{2}}{\frac{1}{8} + \frac{1}{2}} = \frac{4}{5}$

The posterior probability that the consultand's mother is heterozygous is $\frac{1}{5}$. The probability that the consultand is heterozygous is $\frac{1}{10}$ and the probability that a son of hers will inherit muscular dystrophy is $\frac{1}{20}$. The chance that a child with muscular dystrophy will be born to her is $\frac{1}{40}$.

Methods have been developed which help to some extent in identifying females who are carriers of the X-linked muscular dystrophy gene. For example, about two-thirds of carrier females (identified by genetic evidence) have a level of a muscle-derived enzyme (creatine phosphokinase, CPK) in the blood which is greater than that found in 95 percent of normal women. If these studies are done in the consultand, one can include this information in the calculation of the posterior probability. For example, if the enzyme levels are normal, the conditional probability, if the consultand is heterozygous, is $\frac{1}{3}$ and if she is not heterozygous is $\frac{19}{20}$.

Yet another situation for application of Bayes' theorem in genetic counseling is illustrated by the case of a woman (see Fig. 6.12b) with a son who has X-linked muscular dystrophy, there being no other affected males in the family. The alternatives are that the woman is heterozygous (with a risk of 50 percent that any son will be affected) or that the affected son has his ailment as the result of fresh mutation occurring in the X chromosome of a single germ cell of the mother (with a risk essentially zero of recurrence in another son).

If the consultand has several normal sons in addition to the one with muscular dystrophy, the chance of her being heterozygous is reduced. Furthermore, if her serum level of CPK is normal, the probability that she is a carrier is further reduced and the predicted risk to a later-born son is less.

For a lethal X-linked disorder such as Duchenne muscular dystrophy (affected males never reproduce), the prior probability of the consultand being heterozygous by reason of fresh mutation (which occurred in the X chromosome she received from either the father or the mother) is 2μ, where μ is the mutation rate in either parent. (For Duchenne muscular dystrophy, the mutation rate has been calculated to be about 5×10^{-5}. See p. 161.) The probability that she inherited the mutant gene from her mother who was a carrier is also 2μ and therefore the total prior probability of her being a carrier is 4μ. The prior probability of her not being a carrier is $1 - 4\mu$, or essentially 1.

The conditional probabilities are based in part on the genetic information (that the consultand has a normal son as well as the affected one) and in part on the biochemical information (that the serum enzyme level is normal). The genetic findings have a conditional probability of $(1/2)^2$, if she is a carrier; of μ if she is not a carrier. The biochemical finding of normal CPK has a conditional probability of $1/3$, if she is a carrier; of $19/20$ if she is not a carrier. The joint probabilities are: if the consultand is a carrier, $4\mu \times (1/2)^2 \times 1/3$, or $\mu/3$; if the consultand is not a carrier, $1 \times \mu \times 19/20$, or $19\mu/20$.

The posterior, or relative, probabilities are $20/77$, or ~ 0.26 (that the consultand is a carrier) and $57/77$, or ~ 0.74 (that the consultand is not a carrier).

The calculation of posterior probabilities for this example is summarized in Table 6.14.

In the example diagrammed in Fig. 6.12*b* the consultand and her mother had no sibs. If each had several unaffected brothers the probability of the consultand's being heterozygous would be reduced and the same is true if she had other unaffected sons.

TABLE 6.14.   *Bayesian Calculations Relevant to Figure 6.12b*

	Consultand heterozygous	Consultand not heterozygous
Prior probability	4μ	~ 1
Conditional probabilities		
Genetic	$(1/2)^2$	μ
Biochemical	$1/3$	$19/20$
Joint probability	$\mu/3$	$19\mu/20$
Posterior probability	$\dfrac{\mu/3}{\mu/3 + 19\mu/20}$	$\dfrac{19\mu/20}{\mu/3 + 19\mu/20}$
	~ 0.26	~ 0.74

Behavior of Multifactorial Traits in Families

Some traits are not determined predominantly by one gene (or a pair of genes at a single locus) but are determined by a considerable number of collaborating genes, each with rather minor effects. *Polygenic* is one designation for such traits. *Multifactorial* is used interchangeably by many geneticists, but may have a worthwhile separate significance since it takes into account the facts that both genetic and nongenetic factors are involved and that each is multiple. Multifactorial traits are usually of a type that is measurable; unlike discontinuous traits, no black-or-white, yes-or-no, affected-or-unaffected classification of multifactorial traits is possible except by artificial and arbitrary designation or unless, as is discussed later, a threshold phenomenon occurs. Because of this characteristic, multifactorial traits are sometimes referred to as *quantitative traits*. Classic examples in man are intelligence and stature. Blood pressure, refractive index of the eye, dermatoglyphic ridge counts, and many others also qualify as multifactorial traits. In fact, because of the superficial level of phenotype analysis with present methods of study, multifactorial traits are far more numerous than monogenic ones.

Polygenic quantitative traits show a Gaussian, bell-shaped, or "normal" curve when one makes a plot of specific measurement on the

FIG. 6.13. *Frequency distribution curve for height of men in southeast England. From Cedric O. Carter,* Human Heredity. Baltimore: Penguin Books, 1962.

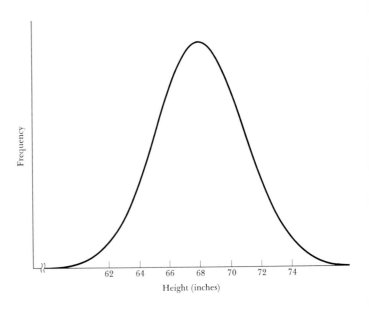

abscissa against number of persons showing that measurement on the ordinate. In Fig. 6.13 is shown the frequency distribution for stature of men living in southeast England. The average height is 68 in. and the standard deviation 2.6 in.

A relatively small number of loci can account for a continuous trait, such as stature. This is illustrated by the hypothetical example shown in Fig. 6.14. Let us assume that stature is determined by two loci on different chromosomes, each with three alleles: One allele (called "–") decreases stature by 2 in., a second (called "0") determines average stature of 68 in., and a third (called "+") increases stature by 2 in. Alleles –, 0, and +, at each locus, have a relative proportion of 1 : 2 : 1. Five classes of gametes produced by each sex have the following frequencies (derived from the so-called Punnett square in Table 6.15) : – –, 1/16; –0, 1/4; 00, 6/16; +0, 1/4; ++, 1/16. Zygotes formed from random union of gametes in these proportions have four stature genes (two at each of the two loci), and the possible combinations of +, 0, and – alleles are nine. The relative frequencies of these are derived from the Punnett square in Table 6.16 and plotted in Fig. 6.14. A satisfactory approximation to a normal curve is obtained. Of course environmental influences blur the separation between groups.

Resemblance among relatives is a useful and important approach for the study of polygenic traits in man. In 1918, R. A. Fisher demonstrated that the findings of the early biometricians do not contradict Mendelian theory but are in full harmony with it and, in fact, can be rationally explained only in terms of Mendelism. He showed that

TABLE 6.15. ***Proportions of Expected Types of Gametes***

Chromosome B \ Chromosome A	*Allele –* 1/4	*Allele 0* 1/2	*Allele +* 1/4
Allele – 1/4	1/16 – –	1/8 – 0	1/16 – +
Allele 0 1/2	1/8 0 –	1/4 0 0	1/8 0 +
Allele + 1/4	1/16 + –	1/8 + 0	1/16 + +

TABLE 6.16. *Proportions of Expected Types of Zygotes*

Male gametes

Female gametes		— — 1/16	— 0 1/4	0 0 6/16	0 + 1/4	+ + 1/16
	— — 1/16	60 in. 1/256	62 in. 4/256	64 in. 6/256	66 in. 1/64	68 in. 1/256
	— 0 1/4	62 in. 4/256	64 in. 1/16	66 in. 6/64	68 in. 1/16	70 in. 1/64
	0 0 6/16	64 in. 6/256	66 in. 6/64	68 in. 36/256	70 in. 6/64	72 in. 6/256
	0 + 1/4	66 in. 1/64	68 in. 1/16	70 in. 6/64	72 in. 1/16	74 in. 4/256
	+ + 1/16	68 in. 1/256	70 in. 1/64	72 in. 6/256	74 in. 4/256	76 in. 1/256

the expected degree of resemblance among relatives, given polygenic inheritance and certain assumptions, is a simple mathematical expression. The regressions of child on parent, or parent on child, or sib on sib, or that of a more remote relative on the proband are equal to the number of genes in common. The expression *genes in common* refers to those genes derived from a common ancestor. The proportion of genes in common, and therefore the regression (Galton's term), for a number of relationships is shown in Table 6.17. The reader has already been introduced to this concept under a different name, the coefficient of relationship, symbolized by r (p. 147).

The correlation coefficient is equivalent to the regression when two regressions, such as child-on-parent and parent-on-child, are equal. Since the mutual regressions are the same in each of the examples shown in Table 6.17, the correlation coefficients are equivalent to the regressions. (The correlation coefficient is the square root of the product of the two regressions, that is, the regression of the first value on

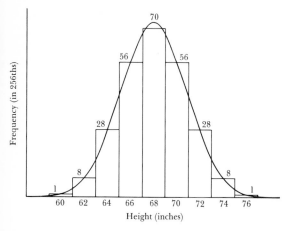

FIG. 6.14. *Distribution curve for stature, assuming that only two loci, each with three alleles with properties and frequencies stated in the text, determine this characteristic.*

the second and the regression of the second value on the first: $\sqrt{r_1 \times r_2}$.)

The conditions under which the theoretical values for intrafamilial resémblances are realized are: (1) that inheritance alone is involved, (2) that the gene pairs do not display dominance or recessiveness, that is, that the heterozygote is intermediate between the two homozygotes, and (3) that values for husband and wife are not correlated, that is, that mating is random and not assortative, as far as the trait under study is concerned.

Of all the quantitative traits that have been studied adequately, fingerprint ridge counts provide one of the best examples of agreement between the expected and observed results (see Fig. 6.15 and Table

TABLE 6.17. *Proportion of Genes in Common*

Relationship to proband	Proportion of genes in common
Parent, child, sib	½
Monozygotic twin	1
Dizygotic twin	½
Grandparent, grandchild, uncle, aunt, nephew, niece, half sib	¼
First cousin	⅛
Second cousin	¹⁄₃₂

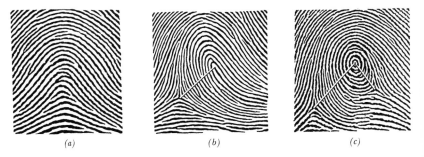

FIG. 6.15. *Examples of three basic types of fingerprint patterns illustrating how ridge counts are made.* (a) *Arch (no triradius). The ridge count is zero.* (b) *Loop (one triradius, on the left where three ridges meet). A line has been drawn from the point of the triradius to the center of the loop. The ridge count, 13 in this case, is the number of ridges cutting the line.* (c) *Whorl (two triradii). The total ridge count is 25. Redrawn from S. B. Holt,* Brit. Med. Bull., 17 (1961), 247.

6.18). The total number of ridges from all ten fingers is taken as the quantitative trait for study. One great advantage is that the dermal ridges are determined rather early in fetal life with no change thereafter, so that no troublesome allowances for change with age are necessary. The frequency distribution for ridge counts in a population approaches a normal bell-shaped curve, as one would expect of a polygenic trait; and, as would probably not be the case if a single gene pair were involved, the curve is unimodal, not bimodal. That multiple loci, widely distributed through the karyotype, are involved in determining total ridge count is demonstrated by the fact that a majority of chromosomal aberrations influence ridge counts.

Certain conditions are necessary for the simplest theoretical expecta-

TABLE 6.18. *Intrafamilial Correlations for Ridge Counts*

Relationship	Correlation coefficient	
	Observed	*Theoretical*
Parent-child	0.48	0.50
Mother-child	0.48 ± 0.04	0.50
Father-child	0.49 ± 0.04	0.50
Parent-parent	0.05 ± 0.07	0
Mid-parent-child	0.66 ± 0.03	0.70
Sib-sib	0.50 ± 0.04	0.50
Monozygotic twin-twin	0.95 ± 0.01	1.00
Dizygotic twin-twin	0.49 ± 0.08	0.50

tions to be realized. There is no assortative mating for ridge counts since the parent-parent correlation is essentially zero. Dominance effects can be tested by using the regression of child on mid-parent (the mean measurement of the two parents). The theoretical regression is 1, since all the child's genes come from the two parents and the mid-parent measurement should be an accurate mean estimate if there is no dominance. The opposite regression, mid-parent on child, is theoretically 1/2 because only half the genes of the two parents are identical to those of the child. The correlation coefficient (defined on p. 156) is, then, $1/\sqrt{2}$, or 0.71. As Table 6.18 indicates, the observed value departs only moderately from that expected, and there is, therefore, little dominance effect.

Stature and intelligence are polygenic in their genetic determination but have complicating features because of assortative mating, dominance, and environmental factors.

References

Carter, C. O., *Human Heredity*. Baltimore: Penguin Books, 1962.

Holt, S. B., "Quantitative Genetics of Fingerprint Patterns," *Brit. Med. Bull.,* *17* (1961), 247–50.

Li, C. C., *Human Genetics*. New York: McGraw-Hill Book Company, 1961.

Maynard-Smith, Sheila, Lionel S. Penrose, and C. A. B. Smith, *Mathematical Tables for Research Workers in Human Genetics*. Boston: Little, Brown & Co., 1962.

Race, R. R., and R. Sanger: *Blood Groups in Man,* 5th ed. Philadelphia: F. A. Davis Co., 1968.

Roberts, J. A. F., "Multifactorial Inheritance in Relation to Human Traits," *Brit. Med. Bull., 17* (1961), 241–46.

Snyder, L. H., "A Table to Determine the Proportion of Recessives to Be Expected in Various Matings Involving a Unit Character," *Genetics, 19* (1934), 1–17.

Genes in Populations

The Hardy–Weinberg principle states that genotype frequencies remain constant from generation to generation unless certain specific disturbing factors are introduced. The factors that disturb the Hardy–Weinberg equilibrium are nonrandom (or assortative) mating, mutation, selection, drift, and gene flow.

It should be clear from study of the algebraic example in Table 6.1 that nonrandom mating (for example, homozygous recessive persons marrying only other homozygous recessive persons) will disturb the Hardy–Weinberg equilibrium and increase the proportion of homozygotes at the expense of heterozygotes.

Mutation

Mutation (for example, from *A* to *a*, occurring during gametogenesis in the parent generation) will change at least slightly the genotype frequencies in the offspring generation (provided, of course, that the mutation is present in a cell that is subsequently passed to the offspring generation). Mutation is a relatively rare event, but it nonetheless provides the raw material of evolution; it provides the basis on which selection, with the collaboration of other factors discussed here, molds the genetic constitution of the species.

The frequency of mutation—the mutation rate—can be estimated in man for some genes, although at best the estimates are only approximations. For dominant genes both direct and indirect methods are available.

The direct method involves ascertaining all cases of a given disorder and determining which of the cases are sporadic, that is, are born of two unaffected parents. The number of sporadic cases is the numerator of the expression for the frequency of mutant individuals; the denominator is the number of births in the period of study. For example, Mørch found 10 achondroplastic dwarfs in 94,075 births in an obstetrical hospital in Copenhagen. Of the 10 cases of achondroplasia, 8 were alleged to have had unaffected parents. The frequency of mutant persons was, then, 1 in about 12,000 births. Since two gametes formed each individual, and since the mutation could have occurred in either the paternal or the maternal gamete, it is necessary to multiply the denominator by 2 to obtain the mutation rate in terms of mutations per gamete per generation; thus, the estimated mutation rate becomes 1 in 24,000, or about 4×10^{-5}. The algebraic expression for the mutation rate (μ), given the number of sporadic cases (n) and the total number of births (N) is

$$\mu = \frac{n}{2N}$$

Several difficulties make the mutation rates determined by the method above no more than "guesstimates." Illegitimacy (or better, nonpaternity), the true father being an affected person, introduces inaccuracy in the labeling of cases as sporadic. So also does mild expression of the gene in an affected parent such that both parents are considered unaffected. Furthermore, the condition under study may not be a single entity having all cases determined by a mutation at one locus. The phenotype may be an environmentally induced phenocopy. Genetic mimics, that is, the same phenotype determined by mutation at more than one locus, may be involved; however, expressing the mutation rate in terms of gametes per generation, rather than in terms of loci, avoids this difficulty. Some of the apparently sporadic dominant cases may in fact be homozygotes for a simulating recessive disorder. In the example of achondroplastic dwarfism used above there is real reason to suspect that heterogeneity exists and that the estimate derived is too high. Finally, total ascertainment is difficult to achieve.

The indirect method is based on an assumption of equilibrium between mutation (which is adding mutant genes to the gene pool) and negative selection (which is removing the mutant genes from the pool). For social and biological reasons, achondroplastic dwarfs reproduce at a much reduced level as compared with the average. The reproductive fitness has been estimated at about 0.20, when the general average is 1.0. This means that achondroplastic dwarfs as a group have only one-fifth as many children as do normal people. Furthermore it means that $1 - 0.20$, or 0.80 of cases of achondroplasia in the next generation arise by new mutation. Mørch found 10 achondroplastic dwarfs in 94,000 births, and presumably 8 of these arose by new mutation. The 94,000 births represent 188,000 gametes; thus, a mutation rate of 8/188,000, or about 1 in 24,000 gametes per generation is calculated. The general expression is

$$\mu = \frac{(1 - f)\, n}{2N}$$

where f is relative fitness, n is the number of cases, and N is the total number of births.

The indirect method has some of the same difficulties as the direct method, for example, the uncertainty of complete ascertainment and of homogeneity of the phenotype. In addition, it is often difficult to be confident of the estimate of fitness, especially in regard to the average number of children in the general population, a value that is used for comparison. If the average number of children of unaffected sibs of cases is used, errors may arise since the normal members of the family may have either more or fewer children for reasons related to the presence of the particular gene in the family.

The indirect method can be used for estimating the mutation rate for X-linked traits, most of which are recessive in the female. Only affected males are ascertained; in the male the mutant allele behaves as though it were dominant. The denominator is multiplied by 3 rather than by 2 to arrive at the estimate per gamete, since the mutation can occur in either gamete of the female or in the X-bearing gamete of the male. Another difficulty in estimating the mutation rate of X-borne genes is deciding whether the mutation in a sporadically-affected male occurred in a gamete of his mother or whether it occurred in a gamete of one of her parents, or even earlier ancestors, and by chance escaped previous detection.

Mutation rates have been estimated for a considerable number of rare pathologic traits of man. Most estimates are of the order of 10^{-5}, that is, 1 in 100,000 gametes per generation.

For autosomal recessive traits, only the indirect method is applicable. Again the question is what mutation rate is necessary to maintain equilibrium (constant gene frequency) in the face of a certain loss of genes through negative selection. Again most of the difficulties enumerated earlier are encountered; in addition, it is very likely that the given recessive gene in the heterozygous state is not entirely neutral but has either a deleterious or an advantageous effect on reproductive fitness. If the heterozygote is at an advantage, a lower mutation rate suffices to replace the genes lost in the homozygote. This selective effect in the heterozygote is difficult to measure, however.

Cystic fibrosis of the pancreas is the most frequent lethal simply inherited disease of childhood, occurring about once in every 3,000 white births in the United States; very few of the persons affected with it reproduce. The evidence for autosomal recessive inheritance of this disease is convincing. To maintain equilibrium between genes added by new mutation and genes lost by early death of homozygotes, a mutation rate of 1 in 1,500 gametes per generation is required. Such an unprecedented estimate is regarded with suspicion, and alternative possibilities are sought. One such possibility is the existence of many loci at any one of which mutation can result in the same phenotype. But even if there are 10 such loci, an average mutation rate of about 1 in 15,000 would be required. A more likely possibility is that the heterozygote enjoys an advantage over the homozygous normal, but as yet there is no evidence that the heterozygote is indeed at an advantage and no proven mechanism by which the heterozygote might be at an advantage.

Another question of interest is whether mutation occurs with equal frequency in spermatogenesis and oogenesis. The time course and other aspects of gametogenesis differ noticeably in the two sexes (see Chap. 2); hence vulnerability to mutation might also differ. In humans the relative frequency of mutation in the male and female can be estimated by an analysis of X-linked traits. If the mutation rate is higher in the female, a disproportionate number of cases in a series will be sporadic, since only the X chromosome given to the son by the mother carries the mutation. If the mutation rate is higher in spermatogenesis, a disproportionate number of cases will be of the familial type, since the one X chromosome of the male bearing the mutation will be transmitted to a daughter who, as a carrier, may have multiple affected sons. Data on hemophilia and on X-linked muscular dystrophy give no clear evidence of a sex difference in mutation rate.

There is evidence that the age of the parent makes a difference in mutation rate. In achondroplastic dwarfism and in certain other conditions the fathers of sporadic cases (that presumably resulted from new mutation) are appreciably older than the average. Correlation of the secondary sex ratio with the age of the maternal grandfather at conception of the mother has been proposed as an approach for determining the effect of age on mutation rate. The secondary sex ratio (ratio at birth) is about 106 males to 100 females. If X-linked recessive mutations, some of them lethal *in utero*, occur with increased frequency in older men, then these might be transmitted to their daughters and segregate in their grandsons, producing a reduction in the ratio.

It is quite certain that most persons, and perhaps all people, carry at least one or two highly deleterious "recessive genes" in the heterozygous state. This follows directly from the calculation of the frequency of heterozygotes of the numerous autosomal recessive disorders that are

known, and is confirmed by investigations of the results of consanguineous marriages. Such studies have led to the concept of "lethal
equivalents." For example, it is estimated that man carries from 3 to 8
lethal equivalents—that is, many genes, each with slightly deleterious
effects, which are the equivalent of 3 to 8 recessive genes that in the
homozygous state would result in death before reproduction.

Causes of Mutation

Definable causes of an increased mutation rate are ionizing
radiation, chemicals, and heat.

Mutations in the general sense are of two types: point mutations and
gross chromosomal changes. The latter is discussed in Chapter 2.
The fact that gross chromosomal aberrations have been observed in
man following exposure to ionizing radiation and nitrogen mustard is
strong evidence that these agents also produce point mutations.

It has been estimated that the total gonadal dosage of X ray (mainly
diagnostic) received by persons in the United States before completion
of the reproductive span averages 3.0 r (roentgens) and is probably not
less than 2.0 r. Fallout from the testing of atomic weapons is approximately 0.1 r. Awareness of the genetic and other risks involved has led
to intense surveillance on the part of the medical and paramedical professions, and elimination of many diagnostic procedures, such as fluoros-
copy and X rays for pelvic size in pregnant women, that represented
heavy exposures.

The mutagenic effect of many chemicals has been proved in other
species. One of these chemicals, nitrogen mustard (or related substances), is used in the treatment of certain malignant neoplastic
diseases of man. Their use is, of course, justified without particular
concern for the mutagenic risks because of the seriousness of the disease
being treated. Furthermore, these individuals usually do not have any
children after such treatment. Many other chemicals in our food or
medical therapeutic armamentarium have mutagenic effects in other
species; their effects in man, however, have not been demonstrated.
Caffeine (trimethylxanthine) and theobromine (dimethylxanthine)
are structurally related to adenine and guanine, two of the four DNA
bases that by various combinations, probably in triplets, encode the
genetic information. Evidence for a mutagenic effect of caffeine is available in *Escherichia coli* and in *Drosophila melanogaster;* however, experiments in mice have thus far failed to demonstrate a mutagenic
effect.

Heat increases the "spontaneous" mutation rate. Measurements of
scrotal temperature in men wearing pants and nude showed an appreciably higher temperature in the former case. Calculations based
on the temperature effect on mutation rate suggest that the wearing of

trousers may be an important factor; some have recommended that kilts would be less mutagenic.

Selection

Selection is another factor that disturbs the Hardy–Weinberg equilibrium by resulting in gene frequencies in the offspring generation different from those in the parent generation. Fitness, in the biological or Darwinian sense, is defined in terms of the contribution made to the genes of the succeeding generation. Selection is differential fitness according to the genotype of the organism, and may operate on both the haploid gamete and the diploid organism. It can furthermore have its effects at any stage from the zygote to the adult individual who has not yet completed the reproductive period. The critical matter is whether reproduction occurs so that the gene is represented in the succeeding generation.

For the purposes of analysis, selection (and its complementary parameter, fitness) can be divided into survivance selection (focusing on the factors that determine whether the organism survives to the time of reproduction) and reproductive selection (focusing on the factors that determine efficiency of reproduction).

On p. 51, it was pointed out that the more severely an autosomal dominant condition interferes with reproduction, the larger is the proportion of ascertained families which have only a single affected person. Figure 7.1a shows the types of families one would expect to find in a collection of families with at least one member affected by an autosomal dominant trait, bearers of which have a fitness half that of the general population—that is, on the average affected persons have only half as many children as do unaffected persons. Assuming equilibrium, one concludes that about half the families will have but one case resulting from new mutation that balances the loss of the mutant gene through negative selection (reduced fitness). Figure 7.1b shows the types of families one would expect in a collection of families ascertained through a male affected by a rare X-linked recessive trait if the males never reproduce and if the fitness of carrier females is no different from that of the general population.

R. A. Fisher pointed out that a dominant mutation that has no selective advantage or disadvantage still has a better than 1 percent chance of survival after 127 generations. Perhaps the longest documented survival of a dominant gene in man is the case of transmission through more than 2000 years of a form of polydactyly (six fingers). This period approximates the 127 generations of Fisher's estimate.

In heterozygotes, gametes carrying one allele may be produced in preference to those carrying the other (so-called meiotic drive). Gametes carrying a particular gene may be at an advantage or disad-

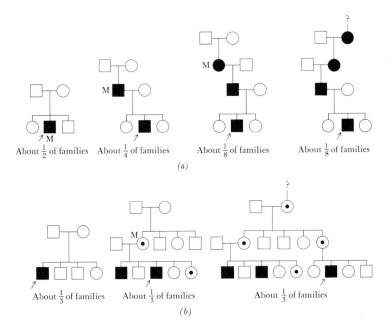

FIG. 7.1. (a) *Types of pedigrees in a collection of families with at least one member affected by an autosomal dominant trait that has a fitness of ½.* (b) *Types of pedigrees in a collection of families wth at least one male affected by an X-linked recessive trait that has a fitness of zero. After C. O. Carter,* Human Heredity *(Baltimore: Penguin, 1962).*

vantage relative to those carrying an allele. Some evidence concerning the occurrence of these phenomena in man has become available in recent years.

The zygote or fetus of a particular genotype may likewise fare better or worse than those of other genotypes. Fetuses heterozygous for Rhesus or some other blood groups are, on the average, at a disadvantage because of the ill effects of materno-fetal incompatibility. Any of the genes causing disorders or predisposition to disorders that lead to death in the years before reproduction obviously place their bearer at a strong selective disadvantage.

One of the clearest examples of selective advantage of a particular genotype in a particular environment is the now familiar one of sickling. The heterozygote enjoys an advantage in a malarious environment. There are at least four indications that malaria has played this role in selection: (1) Geographic distributions of malaria and of high gene S frequency showed a close correlation in Africa. (2) Counts of malarial parasites in the blood are lower in *Hb SA* persons than in *Hb AA* persons. (3) *Hb SA* persons are probably less susceptible than *Hb*

AA persons to experimental malaria. (4) Mortality from cerebral malaria is less in *Hb SA* persons than in *Hb AA* persons. Only item 4 represents direct evidence of selection.

The mechanism of this advantage which the *SA* person enjoys vis-à-vis malaria is not known with certainty. Possibly parasitized red cells become sickled and because of their misshape are removed from the circulation and disposed of more rapidly than would otherwise occur. This mechanism is probably especially important in the early years of life before immunization has had time to develop to the fullest. Pregnant mothers may also be less likely to get the malarial lesions of the placenta that cause abortion, again because the parasitized erythrocytes are removed from the circulation. Whatever the mechanism of the advantage, it prevails only in the face of the malignant form (the falciparum type) of malaria. Hemoglobin C also confers an advantage with reference to malaria. Since Hb C disease (the homozygous state) is not as virulent a disease as Hb S disease, selection against the homozygote is not as rigorous, and less selective advantage on the part of the heterozygote is required to maintain relatively high gene frequencies.

In a nonmalarious environment the sickle heterozygote is actually at a disadvantage. Rupture of the spleen is, for example, likely to occur when the *SA* individual goes to a high altitude in airplane flight. This illustrates an important point: a genotype advantageous in one environment may be quite disadvantageous in another.

In some populations sickle hemoglobin is a polymorphic trait since it is found in a frequency clearly not explicable on the basis of recurrent mutation. Furthermore it probably is a balanced polymorphism, or was under malarious conditions which existed in many parts of the world up until the rather recent past. The loss of *S* genes in the lethal *SS* state was counterbalanced by the greater fitness of the *AS* heterozygote as compared with the *AA* homozygote. As is shown in Table 7.1, if fitness of the *AA* persons is on the average only one-third that of *AS* persons, the frequency of the *S* gene would remain constant at 0.40.

The question is often raised as to whether selection has not been greatly relaxed today as compared with some centuries ago. Certainly the efforts of modern medicine today keep alive individuals who, in an earlier and more precarious setting, would have been easy prey. Undoubtedly the negative selection acting on the gene or genes responsible for conditions such as diabetes, cleft palate, cataract, and retinoblastoma has been appreciably relaxed. Conversely, however, unknown selective factors that, operating in the past, effected a present high frequency of diabetes genes may no longer be at work.

At any rate, much opportunity for selection remains today, and in some ways selection pressure is stronger than in the past. It is estimated that close to half of all zygotes never reproduce. About 15 percent are lost before birth, 3 percent are stillborn, 2 percent die in the neonatal period, 3 percent die before maturity, 20 percent never marry, and 10

TABLE 7.1. *Balanced Polymorphism*

Parent population	Relative fitness		Gametic contribution	
			A	S
AA	36	1/3	24	0
AS	48	1	48	48
SS	16	0	0	0
			72	48
			$p = 0.60$	$q = 0.40$

$$\text{OFFSPRING POPULATION:} \quad \begin{array}{ll} \text{AA} & p^2 = 36 \\ \text{AS} & 2pq = 36 \\ \text{SS} & q^2 = 16 \end{array}$$

percent of those who marry remain childless. Thus, there is considerable room for selection. Genotypic differences that favor survival to reproduction at any of these stages will enjoy an advantage.

The advantage or disadvantage in the race for reproduction can be a rather minor one; yet when applied to large groups of people, and when operating over many generations, the impact on the genetic constitution of man is great. For example, in the case of cystic fibrosis that was cited earlier, an advantage of the heterozygote of only about 1 percent would suffice to maintain the high gene frequency without any mutation. (A 1 percent advantage means that the heterozygotes as a group have 1 percent more children than the homozygous normals.) In part this is due to the large size of the heterozygous population.

The significance of association (for example, blood group and disease association) is the influence it may have had on the genetic constitution of man through selection. It is difficult to believe that the impressive associations between peptic ulcer and blood group O and between peptic ulcer and nonsecretion have had much effect on the world-wide distribution of ABO blood groups. Peptic ulcer, a disease of adulthood, probably does not interfere appreciably with reproduction. On the other hand, it has been suggested that differences in ABO blood type may be related to resistance to infectious diseases such as smallpox and plague, which have been terribly devastating in the past, and that ABO frequencies observed today are the result of selection working through this mechanism. The smallpox virus may bear antigenic similarity to blood group A antigen; persons of blood group B or O, having natural antibody against blood group A, may be more resistant to smallpox. A similar line of reasoning has been applied to plague, the causative organism of which (*Pasteurella pestis*) bears antigenic similarities to H blood-group substance. Correlations of ABO blood-group frequencies with known epidemics of the past and contemporary observations on the severity and outcome of smallpox and plague according to blood type are the main approaches for testing this attractive hypothesis.

Geneticists today do not suppose that any trait, for example, the blood groups, is completely neutral as far as selection is concerned. Selection is thought to be the main reason for differences in gene frequencies in different populations, although drift and gene flow are also important; however, their relative importance is sometimes difficult to evaluate.

Tuberculosis is to a considerable extent a disease of urbanization. It killed large numbers of Europeans in past centuries but had already begun to decline sharply before the advent of effective drugs. Improved hygiene does not adequately account for this decline. A more likely explanation is the rise of more resistant stock through natural selection. The pronounced vulnerability of some groups, such as the Eskimos, American Indians, and Negroes who have not passed through this screen of selection, supports this assertion. The effective treatment of tuberculosis will presumably result in a loss of genetic resistance because of relaxation of selection. However, man will be fitter in other respects. If the genes for resistance to tuberculosis had caused increased fitness in the absence of tuberculosis, then they would have already become frequent in populations. Thus, in the absence of tuberculosis they either make for reduced fitness or are relatively neutral. At any rate, it follows that the drug treatment of tuberculosis is not dysgenic and is probably eugenic.

Drift

Drift, originally called random genetic drift by Sewall Wright, who developed the concept, occurs in small populations or isolates that form for geographic, religious, or social reasons. Although isolates have all but disappeared today because of improved means of transportation and communication, increase in the world's population, and urbanization, man was a relatively rare mammal that existed in small groups, in the not so remote past.

In statistics, the standard deviation in small experiments is greater than in large experiments. If a small number of animals has been tested in an experiment, the "confidence limits" on the value determined as the average for the variable tested are much wider than if a large number of animals is used. Drift is a comparable matter: in small populations the frequency of one gene may by chance rise to high proportions. In a small religious isolate group, the Dunkers of Pennsylvania, the frequency of the genes for blood groups A, B, and O was found to be quite different from those of the parent population in Germany from which the Dunkers were derived. Effects of this type might have been stronger in earlier times when the total human population of the earth was small and men lived in small groups, especially on islands or in other geographically isolated areas.

The founder effect, like drift, is based on randomness or chance events. If an unpopulated area was colonized by a few married couples, by chance differing widely from the average of the parent population in the genes they carry, for example, for various blood groups, the descendant population might differ markedly from the parent population.

An example of founder effect is the unprecedented frequency in the Old Order Amish of Lancaster County, Pennsylvania, of the gene which in homozygous state causes a particular form of dwarfism accompanied by extra fingers (polydactyly). Almost all 8,000 persons in this isolate include among their ancestors three couples who came to America before 1770. Although dwarfism-polydactyly, a simple recessive (see Table 6.8, p. 141), is so rare that scarcely more than 50 cases had hitherto been reported in the medical literature, over 61 definite cases have been found among the Amish of Lancaster County; about 13 percent of persons in this group are heterozygous for the gene. By chance, one of the "founders" must have been a carrier of this gene, which is very rare in the general population. Amish groups living elsewhere than in Pennsylvania, for example, in Ohio and Indiana, had other founding fathers and do not have this abnormality. (To some extent, random genetic drift must have contributed to the high frequency of the six-fingered dwarfism gene in the Lancaster County Amish in the face of its detrimental nature in the homozygote. Almost certainly 13 percent of the founding group was not heterozygous for the gene. Indeed, probably only one founder out of about 200 was heterozygous.)

Note that random genetic drift represents decrease in the originally existing variability because of chance gametic choices, whereas founder effect is poverty of genetic variability existing from the beginning because of chance zygotic choices.

It is often difficult to evaluate the relative importance of selection and drift in determining gene frequencies observed today. For example, there is a high frequency of blood group O (almost 100 percent) in South American and many North American Indians, whereas Oriental peoples from whom they were derived have a rather low frequency of blood group O. Whether the shift to high O in American Indians resulted from selection or from drift is unknown. Drift probably was responsible for certain atypical frequencies such as the high frequency of blood group A (up to 80 percent) in Blackfoot and Blood Indians, as compared with 2 percent in the Ute Indians.

Gene Flow

Gene flow is the change in the genetic constitution of populations as a result of addition of new genes by migration, contact with invading armies, and miscegenation. Selective migration can influence the make-up of both the donor and the recipient population.

The distribution of ABO blood groups illustrates the influence of migration on genetic constitution. In Great Britain there is a cline from high A-gene frequency in the south of England to high O-gene frequency in Scotland (Fig. 7.2a). This cline is thought to be due to the progressive northward retreat of aboriginal peoples with high O frequency before the pressures of more recent immigrants with high A frequency from the Continent. In Ireland, too, the varying mixtures of the original Irish on the one hand and the Norman and English in-

FIG. 7.2a. *Cline for blood-group A gene in United Kingdom, Ireland, and Iceland. Redrawn from Mourant et al. (1958).*

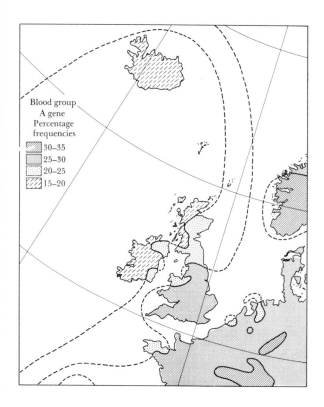

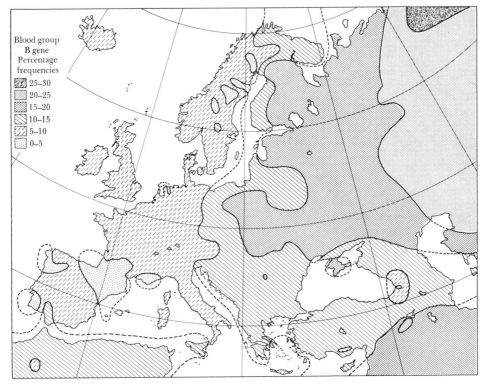

FIG. 7.2b. *Cline for blood-group B gene on Eurasian continent. Redrawn from Mourant et al. (1958).*

vaders on the other are shown by the cline from high O frequency in the west to relatively high A frequency in the east of the island.

A similar cline across Asia and Europe from high B and low A frequencies in the East and to low B and high A frequencies in the West (Fig. 7.2b) has been ascribed to effects of the Tartar-Mongol invasions of 500–1500 A.D. It is found that descendants of refugees who fled into the mountain strongholds of the Caucasus have low blood-group B frequency, and retain the social and linguistic traits of their forebears.

Africans indigenous to the area south of the Sahara have certain blood-group peculiarities that distinguish them from other peoples. One of these features is high Rh type cDe (R_0). The frequencies of the blood-group genes seem relatively stable, thus permitting an estimate of the degree of miscegenation of Caucasians and Africans in the United States. For social reasons gene flow has been almost exclusively from Caucasian to African. Using the Rhesus blood group, one arrives at an estimate that on the average the American Negro is about 30 percent Caucasian, that is, about 30 percent of the total gene pool carried by American Negroes is of Caucasian contribution. Using other polymorphisms with strikingly different frequencies in Africans and Europeans, almost identical estimates of white–Negro admixture have been obtained.

Inbreeding

Consanguinity per se does not affect gene frequency in a population. It is a popular misconception that inbred groups suffer a "building-up of bad genes." Consanguinity changes *genotype* frequencies, but not gene frequencies, by increasing the proportion of homozygotes. If the phenotype of a homozygote is deleterious, inbreeding will actually cause a decline in the frequency of that gene. This is the basis of the concept of "inbreeding bottleneck"; it is possible that through close marriage over many centuries a population might reduce the average number of rare recessive genes carried by each individual as compared with an outbred population.

Man's Genetic Diversity

Polymorphism (p. 88) for many genes of man has been recognized. All groups of man are polymorphic for 12 or more blood-group systems controlled by separate loci and "normal" variation at many other loci has been identified. What, on the average, is the degree of heterozygosity in the human species? Analysis of the experiences with blood groups suggested that man may be polymorphic at as many as 16 percent of his genetic loci. In another experiment, 10 enzymes were studied by electrophoresis in about 100 persons. Polymorphism was discovered in 3 of the 10 enzymes. These two experiences suggest that man has a high order of genetic diversity of a "normal" type. Selection must be a leading force in the establishment and maintenance of this presumably advantageous diversity, but the details are totally unknown.

Terms

Certain terms used in population genetics must be clearly defined. *Incidence* and *prevalence* have acquired rather specific meanings, mainly in studies of the epidemiology of infectious diseases. *Incidence* refers to the number of new cases that develop in a certain period of time. In human genetics it would be appropriate to speak of the incidence of Down's syndrome as 1 in 500 births since a period of time, that required for 500 births to occur, is implicit. *Prevalence* refers to frequency of "cases" found in a population at any one point in time. It is appropriate to speak of the prevalence of Down's syndrome in a given population in which total ascertainment has been achieved. No lapse of time is implied; the enumeration is at one point in time. The main confusion involves the term *incidence*. To avoid such confusion the term *frequency* has much to recommend it. It should, however, be made clear whether phenotype frequency or gene frequency is referred to.

References

Ehrenberg, L., G. von Ehrenstein, and A. Hedgran, "Gonad Temperature and Spontaneous Mutation-Rate in Man," *Nature, 180* (1957) , 1433–34.

Glass, Bentley, "Genetic Changes in Populations, Especially Those Due to Gene Flow and Genetic Drift," *Advan. Genet., 6* (1954) , 95–139.

Hardy, G. H., "Mendelian Proportions in a Mixed Population," *Science, 28* (1908) , 49–50.

Lewontin, R. C.: "An Estimate of Average Heterozygosity in Man," *Am. J. Human Genet., 19* (1967) , 681–85.

Manoiloff, E. O., "A Rare Case of Hereditary Hexadactylism," *Am. J. Phys. Anthrop., 15* (1931) , 503–08.

McKusick, V. A., "The Distribution of Certain Genes in the Old Order Amish," *Cold Spring Harbor Symp. Quant. Biol., 29* (1964) , 99–114.

Mourant, A. E., and others, *The ABO Blood Groups.* Oxford, England: Blackwell Scientific Publications, 1958.

Neel, J. V., and W. J. Schull, "On Some Trends in Understanding the Genetics of Man," *Perspectives in Biol. & Med., 11* (1968) , 565–602.

Woolf, C. M., and F. C. Dukepoo, "Hopi Indians, Inbreeding, and Albinism," *Science, 164* (1969) , 30–37. The relationship between social structure and genetic structure of a population is illustrated.

Eight

Genes and Evolution

Separation of a discussion of "genes and evolution" from "genes in populations" is entirely arbitrary. Evolution in simplest terms is the change in the genetic constitution of an organism. The important factors in evolution—those influencing gene frequencies—were discussed in Chap. 7. Several additional topics, however, are worthy of discussion.

R. A. Fisher pointed out that grave genetic defects have not always been recessive to the wild type as now seems to be the case for a majority of conditions. He postulated that through the accumulation of genetic modifiers the effects of genes in the heterozygote were gradually mollified to the point that their action is recognizable only, or almost only, in the homozygote. It is not difficult to imagine that when a deleterious dominant mutation was occurring repeatedly, modifying genes that reduced the deleterious effects would be favored in selection. Fisher originated the expression "the evolution of dominance" for this phenomenon.

Evolution of the genetic material itself can be deduced from the nature of certain proteins of related structure. Unequal crossing over between homologous chromosomes and subsequent divergence through independent mutation of the duplicate loci is thought to be a frequent mechanism. This phenomenon probably accounts for the findings in the hemoglobins. The locus responsible for the synthesis of delta chains of hemoglobin A_2 is very closely linked to the locus determining beta chains of hemoglobin A, and there are close chemical similarities between the β-polypeptide chain and the δ-polypeptide chain. In fact they differ by only ten amino acids. Hence it is logical to assume that gene duplication and divergent mutation have occurred.

Gene duplication is a mechanism that permits more rapid and more extensive evolution of proteins than can be achieved by mutation alone. Furthermore, it has the advantage that the parent protein, with any beneficial effects it may have, need not be lost.

Unequal crossing over may be intragenic rather than intergenic with the production of a new allele that has some characteristics of one gene and other characteristics of another. The close study of amino acid sequences of certain proteins of man has turned up several instances where this has probably happened.

Parenthetically, it should be noted that the analysis of protein structure is a powerful, although indirect, method for fine-structure genetic analysis in man—genetic analysis at the intragenic level. Intragenic inversions, deletions, shifts, and other changes can, at least theoretically, be identified by this method.

Deductions on the evolution of the several polypeptide chains of the hemoglobins can be made from chemical homologies, that is, the similarities and dissimilarities in amino acid sequence. Hemoglobin probably arose from a relatively simple myoglobin-like protein with one polypeptide chain. The loci determining alpha and beta chains must have diverged from the primordial locus in the remote past. The delta locus probably evolved from the beta locus through gene duplication; evidence for this relatively recent evolution is provided by the facts that Hb A_2 occurs only in primates, that there is about 95 percent amino acid homology between the delta and beta chains, and that the delta and beta loci are closely linked. The gamma locus appears to

FIG. 8.1. *The evolution of the human oxygen-transporting and oxygen-storing proteins, hemoglobin and myoglobin. Based on V. M. Ingram,* **Federation Proc., 21 (1962), 1053–57.**

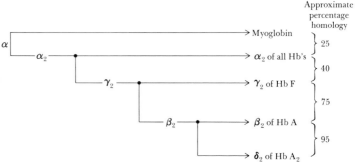

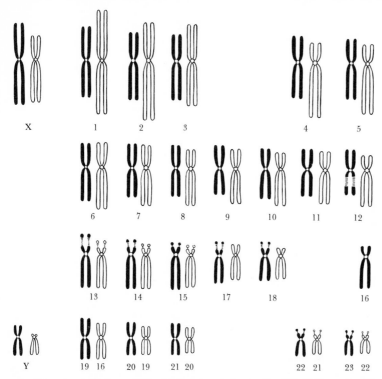

FIG. 8.2. *A comparison of the karyotype of the gorilla (black figures) with that of man (open figures). The gorilla has 48 chromosomes. After J. L. Hamerton et al., Nature, 192 (1961), 225. (Hamerton now concludes that the X chromosome of the gorilla is the fifth or sixth in size, not the largest as shown here.)*

have originated at an intermediate stage. (These relationships are schematized in Fig. 8.1.) Indeed there appear to be at least two separate gamma loci which presumably originated by the process of gene duplication. The fetal hemoglobins produced by the separate loci are not distinguishable by most physicochemical methods, but amino acid sequencing shows that their gamma chains have at least one amino acid difference: at position 136, one type has glycine whereas a second type has alanine.

At the chromosome level as well as at the protein level, the course of evolution can be traced by the study of homologies. As is indicated in Fig. 8.2, there are some striking similarities between the karyotype of man and that of his rather close relative, the gorilla. There is reason to think that homologous chromosomes, for example, the X chromosome of gorilla and man, may have similar genetic content. Linkage and other methods for genetic analysis in man are crude and laborious. Based on the reasonable hypothesis of homology, genetic analysis in man may be furthered by study in other species in which experimental matings are possible.

An interesting hypothesis, which has some support, suggests that man is tetraploid as compared to primitive species. It is suggested, for example, that considerable homology persists between the two pairs of group G chromosomes. In the light of the tetraploid hypothesis it becomes irrelevant to discuss whether Down's syndrome is produced by triploidy of chromosome 21 or chromosome 22. Triploidy of either chromosome may lead to Down's syndrome; there may be two clinically slightly different forms of Down's syndrome which we have not yet learned how to distinguish. The evolutionary advantages of tetraploidy might be similar to those of gene duplication.

The mechanisms leading to balanced polymorphism, such as the relatively high frequency of color blindness in many human populations, are not clearly understood. One plausible explanation for this particular polymorphism is that in a society of hunters the color-blind man was less confused by the camouflaging mimicry that evolved in many animals as a protection against its predators. An opposite view is that color blindness was a disadvantageous trait in a hunting society and that selection against the trait has been relaxed in recent millenniums. The proponents of this hypothesis point to the low frequency of color blindness in aboriginal peoples.

We have seen how a "pathological gene," that for sickle hemoglobin, may be advantageous in some environmental circumstances. Myopia (nearsightedness), a polygenic trait, may have been advantageous in early societies. The young man incapable of hunting because of the visual defect may have been forced into intellectual pursuits that in the long run were more advantageous to society. Shocking as the notion may seem, even death may have a selective advantage: in primitive societies aged persons no longer capable of reproduction or of productive labor are a burden.

Races

Races of man are distinguished by their possession of aggregates of different genes, or the same genes in different frequencies, or both. There are, however, numerous genetic similarities between races, making classification difficult. In fact, no classification satisfactory to all anthropologists has been devised. The standard classification that is supported by differences in blood group frequencies is:

(1) European (Caucasoid)
(2) African (Negroid)
(3) Asiatic (Mongoloid)
(4) American Indian
(5) Australoid

A more elaborate classification with its basis in both geography and physical traits is:

(1) Amerindian
(2) Polynesian—islands of eastern Pacific, New Zealand to Hawaii and Easter Island
(3) Micronesian—islands of western Pacific, from Guam to Marshall and Gilbert Islands
(4) Melanesian—islands of western Pacific, from New Guinea to New Caledonia and Fiji
(5) Australian
(6) Asiatic
(7) Indian—populations of the subcontinent of India
(8) European
(9) African

Even finer classifications—into as many as 34 races—have been suggested. These classifications list, for example, the Bantu, Pigmy, North American Colored, Eskimo, and others as separate races. All races are Mendelian populations that change in time through the influences of the factors discussed earlier. Some of the racial groups in the more detailed classifications have arisen in the last 400 years or less, for example, the North American Colored and the Neo-Hawaiian.

At any rate, man is all one species with no chromosomal differences between the various races and with free interbreeding possible. New subraces may develop in the future through intermixture, but new major races are not likely to evolve unless man is widely scattered into small groups by a nuclear war or other form of cataclysm.

Thus far, studies have failed to reveal evidence for hybrid vigor in interracial crosses in man, but this is perhaps not surprising since man is already extensively outbred.

References

Bryson, V. and H. J. Vogel, eds., *Evolving Genes and Proteins* (Symposium held at Rutgers University) . New York: Academic Press, Inc., 1965.

Chu, E. H. Y., and M. A. Bender, "Cytogenetics of the Primates and Primate Evolution," *Ann. N.Y. Acad. Sci., 102* (1962) , 253–66.

Coon, Carleton S., *The Story of Man,* 2nd ed. New York: Alfred A. Knopf, Inc., 1962.

———, *The Origin of Races.* New York: Alfred A. Knopf, Inc., 1962.

Dobzhansky, Theodosius, *Mankind Evolving: The Evolution of the Human Species.* New Haven: Yale University Press, 1962.

Hamerton, John L., *et. al.* "Somatic Chromosomes of the Gorilla," *Nature, 192* (1961) , 225–28.

Hoagland, Hudson, and R. W. Burhoe, eds., *Evolution and Man's Progress.* New York: Columbia University Press, 1962.

Howells, W., *Mankind in the Making. The Story of Human Evolution* (revised ed.). Garden City, N.Y.: Doubleday & Co., Inc., 1967.

Huntington, Ellsworth, *Mainsprings of Civilization*. New York: John Wiley & Sons, Inc., 1945.

Ingram, Vernon M., *The Hemoglobins in Genetics and Evolution*. New York: Columbia University Press, 1963.

Schroeder, W. A., T. H. J. Huisman, J. R. Shelton, J. B. Shelton, E. F. Kleihauer, A. M. Dozy, and B. Robberson, "Evidence for Multiple Structural Genes for the Gamma Chain of Human Fetal Hemoglobin," *Proc. Natl. Acad. Sci. U.S., 60* (1968), 537–44.

Nine

Genes and Disease

Medical genetics is the aspect of human genetics that is concerned with the relationship between heredity and disease.

The mutant gene should be viewed as an etiological agent in disease, comparable to a bacterium or a virus. One sometimes hears that the aim of a particular research program is to discover the *cause* of hereditary muscular dystrophy. Actually the cause has long been known. The cause of muscular dystrophy is a mutant gene (or several mutant genes, since several forms of hereditary muscular dystrophy can be distinguished). What is sought in such research are the mechanisms, the nature of the gene-determined biochemical defect in each form of muscular dystrophy.

Genetic factors are involved in all diseases. It is difficult, if not impossible, to imagine one in which genetic factors play no role or one in which environmental (or exogenous) factors play no role. Both are involved in determining any phenotype, including the abnormal phenotypes considered diseases. The relative significance of genetic and exogenous factors varies from disease to disease (and perhaps even from case to case). All disease can be viewed as falling on a spectrum (Fig. 9.1) in this regard. Some, like galactosemia and PKU, are near the genetic end (**G**) of the spectrum, but not at the extreme end because the capacity of dietary modifications to ameli-

FIG. 9.1. *A spectrum of disease as to the relative importance of genetic and exogenous factors.*

orate the phenotype indicates a significant exogenous element in these diseases. A disease like tuberculosis is near the exogenous end (**E**) of the spectrum but again not at the extreme end because studies making use of the twin method, interethnic comparisons and animal analogues (see p. 190) indicate that the genetic constitution of the individual plays a significant role in determining susceptibility to the infection and the severity of disease after infection. Other diseases, such as diabetes mellitus and essential hypertension, fall somewhere near the middle of the spectrum.

It is useful to think of all diseases as comprising three classes, as far as genetic factors in their causation are concerned: (1) Some are primarily "caused" by a *single mutant gene* (in heterozygous, homozygous, or hemizygous state depending on the mode of inheritance). These "Mendelizing" disorders are individually rare but in their aggregate constitute a significant body of disease. Huntington's chorea, cystic fibrosis, and hemophilia are examples. (2) Other diseases are "caused" by chromosomal aberrations identifiable by presently existing methods (see Chap. 2). Usually these disorders are not inherited—at least not in the usual sense of the word—but they do involve the genetic material and therefore represent one category of genetic disease. Down's syndrome is an example. (3) Many other diseases are multifactorial in their causation. Both genetic and exogenous factors are multiple and interact in a complex manner to produce a phenotype which is considered a disease. *Polygenic* is a synonym for the genetic part of multifactorial causation. Many relatively common ailments of man are multifactorial in their causation. Essential hypertension and peptic ulcer are probable examples of conditions with multifactorial causation.

This three-way classification is, to a considerable extent, arbitrary. Single gene disorders show modification by both the rest of the genome and the environment, and all of them, as indicated by the spectrum shown in Fig. 9.1, are in a sense multifactorial. Since multiple genes are involved in chromosomal aberrations, one might view them as polygenic or multifactorial. Despite its arbitrariness, the classification is useful, however, because the implications of genetic principles for diagnosis, prognosis, and treatment are somewhat different in the three categories of disease.

William Bateson, who in 1905 first introduced the term *genetics,* defined his term as the science of variation. Although the geneticist focuses principally on gene-determined variation, he must keep all sources of variation in view. The dividing line between normal and abnormal variation is not fixed; indeed no dividing line exists. "Disease" or "dis-order" are not easily defined terms. Variation in one environment or social setting may be considered pathologic, whereas in another it may be advantageous. Keep in mind that variation constitutes a continuum as far as effects on the health and happiness (and survival) of the individual are concerned, and the matters discussed in this chapter are not in principle different from those discussed elsewhere. All variation, not merely pathologic variation, can, for example, be viewed in the three categories: single gene, chromosomal, and multifactorial.

The practice of medicine resolves itself mainly into seeking the answers to three questions: What is wrong? (diagnosis) ; What is going to happen? (prognosis) ; and What can be done? (treatment) . (In addition it is the scientific and social responsibility of the physician to keep in mind a fourth question: Why did it happen?) Diagnosis, prognosis, and treatment provide a useful framework within which to discuss the role of genetics in the practice of medicine.

Diagnosis

In diagnosis, genetic knowledge is useful in recognizing grave internal disease from external clues that are part of a hereditary single-gene syndrome (see p. 86). The principle of *pleiotropism* is used. Also, early manifestations or mild expression (*forme fruste*) of hereditary disease can often be recognized through knowledge of the family history and the usual familial pattern of the genetic disorder in question. Development of relatively easy methods for studying the chromosomes of man has increased greatly our diagnostic abilities.

In some conditions in which the exact mode of inheritance is not yet known, for example, diabetes mellitus, the knowledge that the disorder runs in families and that the particular family in question has several members affected can at least serve the useful function of arousing suspicion that this disorder of carbohydrate metabolism is the basis for otherwise unexplained symptoms in the patient. If the parents of a patient are related (a rare occurrence in this country at present) the pediatrician may be inclined more strongly toward the possibility of an autosomal recessive disorder as the basis of the child's symptoms.

Prognosis

Prognosis in medical genetics has features that distinguish it from prognosis in other fields of medicine. Frequently the question, "What is the outlook?" applies to the unborn offspring of the persons seeking advice, and not to the persons themselves. Once a child is born with a hereditary disorder, the parents often wonder what the risk is that another child will be similarly affected. Prognostication in medical genetics is covered by so-called genetic counseling. The person seeking counsel is often not ill and therefore not a patient; *consultand* is a useful term for the person being counseled.

The first principle of genetic counseling is that there is no substitute for an accurate diagnosis. Genetic heterogeneity (see p. 85) must be kept in mind. The importance of diagnosis is illustrated by the following experience: normal parents had a child born with a disorder diagnosed as achondroplasia with clubbed feet. Inquiry by mail to a geneticist yielded the advice that since achondroplasia is well known to be dominant and since the parents are normal, the child's condition must represent the result of fresh mutation and the risk of recurrence is negligible. The parents had other affected children because in fact the proband had a disorder called diastrophic (meaning twisted or bent) dwarfism which is distinct from true achondroplasia and is inherited as an autosomal recessive.

In the case of many conditions, most of them individually rare, it is possible to state the genetic risk in terms of rather precise probabilities. For example, once a child affected by a clearly autosomal recessive disorder has been born of unaffected parents the risk to any child subsequently born is, of course, one in four. Even if three or more children, all affected, have been born, the risk is still one in four. "Chance has no memory." For any child of a person affected by an autosomal dominant disorder (that has reasonably complete penetrance) the risk is one in two.

In the case of X-linked disorders (most behave as recessives), the problems of genetic counseling are somewhat different. What, for example, is the risk that a sister of a hemophiliac will have a hemophilic son? The risk in such a case varies greatly, depending on whether the hemophilic brother inherited his disease from a carrier mother or whether his disease arose through a new mutation in the X chromosome contributed by his mother. If there are affected maternal uncles, then the answer to the question is one in four. The risk of the sister being a carrier is one in two and the risk of her having a hemophilic son if she is a carrier is again one in two; the risk to a son is, then, the product, one in four.

Obviously, the ability to identify persons heterozygous for a "recessive gene," either X-linked or autosomal, would be of great value in genetic counseling. The methods presently available give, for most conditions, an imperfect separation between the heterozygote and the normal homozygote. If the test—for serum enzymes in the case of the sister of a boy with X-linked muscular dystrophy, or for antihemophilic globulin in the case of the sister of a hemophiliac—yields a value well outside the normal range, then it is possible to state with considerable certainty that the person is a carrier. However, even if the value is in the normal range, one cannot be confident that the person is not a carrier. Most serious conditions with recessive inheritance are rare. It is much more important to be able to identify the carrier of an X-linked recessive than the carrier of an autosomal recessive. A 50 percent chance of being affected is carried by any son of a woman heterozygous for an X-linked condition, regardless of the genotype of his father. On the other hand, a person heterozygous for an autosomal recessive has a chance of having an affected child only if he is married to another heterozygous person (or mutation occurs in the gamete contributed by the spouse).

In Chap. 6 (p. 151), Bayesian methods for deriving the probability that a given woman is heterozygous for a given X-linked gene were described, using the example of X-linked muscular dystrophy.

In many conditions, for example, congenital malformations such as harelip, cleft palate, and congenital heart disease, the role of genetic factors is sufficiently unclear that genetic risks can be estimated only in empiric terms. For example, in a collection of families with normal parents and a single child with harelip, a frequency of 4 percent of harelip may have been found among offspring born subsequent to the affected child. If one parent is also affected, the risk is increased, and so on. Empiric-risk figures undoubtedly exaggerate the risk in some families and underestimate it in others. In this situation, as in so many others in medical genetics, the likelihood of heterogeneity must be kept in mind. One looks for objective features that may make it possible to arrive at a more precise estimate of the genetic risk in some cases. There are, for example, rare forms of cleft palate that have simple Mendelian inheritance. Although the genetics of these congenital malformations is complex and information imprecise, the risk of recurrence is low; the more complex the genetics, the lower the risk.

Often genetic counseling affords relief from worries rather than inciting or aggravating anguish. Usually the person seeking advice thinks the risk is greater than it really is. Persons in hemophilic families may be worried that they can have affected children and are likely to be relieved if pedigree analysis demonstrates that it is unlikely or completely impossible for them to transmit the disorder; the nonhemophilic brother of a hemophiliac cannot transmit hemophilia to his descendants, but he may not know this. A man with pseudoxanthoma elasticum (failing vision and severe disease of blood vessels) inherited as an autosomal recessive is relieved to learn that the risk is negligible that his children by an unrelated wife will be affected.

Another aspect of genetic prognostication is illustrated by Huntington's chorea, a neurologic disorder in which the age at onset of manifestations (abnormal movements and eventual dementia) may vary from the first to the seventh decade. A member of a family with multiple persons affected by this disorder in an autosomal dominant pattern may ask what the risk is that he carries the gene and that manifestations of the disorder will appear later in his life? Unfortunately there are no foolproof early signs of the disease to help in answering the question. A tentative answer is illustrated by this example: The consultand is a 20-year-old man without signs of Huntington's chorea who has an unaffected father aged 50 years. The paternal grandmother died of Huntington's chorea at 70 years of age. The chance that the consultand has the gene is dependent on whether the father has the gene. About 80 percent of cases of Huntington's chorea have become clinically evident by the age of 50 years. Thus, according to the Bayesian principles outlined on p. 151, the prior probability of the father's having the gene is $\frac{1}{2}$ and the conditional probability that if he inherited the gene he has not shown it is $\frac{1}{5}$. The joint probability, of having inherited the disease and appearing normal, is therefore $\frac{1}{10}$. The prior probability of the father's not having the gene is also $\frac{1}{2}$, the conditional probability of not showing the disease is 1, and the joint probability is $\frac{1}{2}$. The posterior or relative probability of the father's having inherited the disease is

$$\frac{\frac{1}{10}}{\frac{1}{2} + \frac{1}{10}} = \frac{1}{6}$$

The estimated probability that the son will be affected is then 1 in 12, or 8.3 percent.

Useful assistance can be provided in predicting the course of the disease in a person who already has symptoms if, in a given family, both the similarity and the range of variability in the manifestation of a mutant gene are taken into account. For example, in different families the form of retinal degeneration called retinitis pigmentosa shows differences in the rate at which the disorder progresses to blindness.

Treatment

Treatment in genetic disease is not as hopeless as it may seem. There is, of course, a difference between treatment and cure; cure is not now possible unless the surgical correction of genetically determined abnormalities can be considered to be a cure. Opportunities for treatment in genetic disease arise in part from the fact that almost all disease is at least to some extent the result of collaboration of environmental and genetic etiologic factors (see p. 181).

There are several forms of therapy for genetic diseases:

(1) *Substrate restriction by elimination diets.* In galactosemia and PKU, galactose and phenylalanine, respectively, cannot be metabolized properly; their accumulation results in pathologic symptoms. Elimination of galactose and phenylalanine from the diet at an early stage can prevent irreversible damage.

(2) *Supply of the substance which cannot be synthesized.* In oroticaciduria there is a genetic defect in the synthesis of uridylic acid and cytidylic acid, both essential to normal pyrimidine metabolism. If uridine is taken orally in adequate amounts, no manifestations of the enzyme defect will develop. Avitaminoses and deficiencies of essential amino acids are not viewed as primarily genetic disorders since all human beings seem to require these dietary elements. However, as compared with some other mammalian species, man has a genetic defect in vitamin C synthesis; "treatment" of this genetic defect is dietary consumption of vitamin C. Thyroid hormone may be administered to treat genetic defects of thyroid hormone synthesis.

(3) *Avoidance of drugs.* Certain antimalarial and other drugs and also the fava bean precipitate hemolysis in persons with the X-linked genetic deficiency of erythrocyte G6PD. Clearly, preventive treatment consists of avoiding the offending agent.

(4) *Elimination from the body.* Hemochromatosis is a hereditary disorder in which, by some mechanism not yet clearly understood, iron accumulates in the body in very large amounts, producing severely deleterious effects on the heart, liver, and pancreas. An effective method for removing iron from the body is repeated venesection.

(5) *Competitive inhibition.* In oxalosis, because of a defect in the degradation of glyoxalate to CO_2 and H_2O, glyoxalate is converted in greatly excessive amounts to oxalate. The administration of sodium hydroxymethane sulfonate, which acts as a substrate for the same enzyme involved in the conversion of glyoxalate to oxalate, may be effective treatment.

(6) *Enzyme induction.* Probably in many inborn errors of metabolism the particular enzyme concerned is formed, but because of mutation is a "warped molecule" with, let us say, only 2 percent of the enzyme activity of its wild-type counterpart. If some method can be found to induce a tenfold increase in the amount of abnormal enzyme produced, although the total enzyme activity is brought only to 20 percent of the normal, it might make a considerable difference in the functioning of the individual. The validity of this approach is demonstrated in a form of hereditary jaundice caused by a defect in the enzyme which catalyzes the chemical conjugation of bile pigment, an essential step in its excretion into the bile. With administration of phenobarbital, jaundice may disappear because of induction of enzyme production by the drug. Some steroid hormones, like cortisone, also have an enzyme-inducing action.

(7) *Enzyme repression.* In oroticaciduria (mentioned earlier), excessive orotic acid is formed and accumulates in the urinary tract. Administration of uridine not only corrects the anemia by supplying a metabolite which is essential to nucleic acid metabolism, but also the synthesis of excessive orotic acid is prevented through feedback repression of early steps in the metabolic pathway.

(8) *Cofactor supplementation.* Some vitamins, such as B6 (pyridoxin), biotin, and B_{12}, serve as cofactors for particular enzymes. In some inborn errors of metabolism it is now known that ingestion of increased amounts of the cofactor will restore the metabolic functions nearly to normal. One interpretation is that the mutant enzyme is impaired in its function mainly because of improper interaction with the cofactor and compensation is possible with increased concentration of the cofactor.

(9) *Replacement of defective tissue.* Kidney transplantation in hereditary cystic disease of the kidney is a potentially feasible form of therapy, but the problems of tissue incompatibility must first be solved.

(10) *Preventive therapy.* Surgical removal of the colon in cases of hereditary polyposis of the colon, a condition notoriously liable to develop into cancer, is an example. Prevention of Rh sensitization has been discussed on p. 116.

(11) *Other forms of surgical therapy.* Removal of the spleen in hereditary spherocytosis corrects the main manifestation, anemia.

Perhaps someday tailor-made, viruslike agents may be used to change the genetic makeup of human individuals by a process similar to transduction in bacterial genetics. This would indeed be cure! However, it will undoubtedly be easier to influence steps between DNA and the protein that it specifies or to alter the quantitative activity of genes, as suggested above, than to alter the structure of DNA itself. For example, if we knew how to reverse the switch from gamma to beta chain (fetal to adult hemoglobin) synthesis, we could cure sickle-cell anemia. Adults with only fetal hemoglobin (a condition called hereditary persistence of fetal hemoglobin) are seemingly normal.

In the area of therapy, genetics also impinges on clinical medicine in connection with the genetic differences in response to drugs—a relatively new area of study in medical genetics called *pharmacogenetics.*

Prevention of malformations by treatment of the pregnant mother who has a chance of bearing an affected child has precedence in animals, although no example in man has been established. In mice a congenital eye anomaly is prevented by treating the pregnant mother with cortisone, and a congenital disturbance of balance is avoided by giving the mother dietary supplements of manganese during pregnancy.

Preventive medicine based on genetic principles can be practiced by discouraging the marriage of two individuals heterozygous for the same recessive gene. This is being done on the largest scale in Italy where schoolchildren are screened for evidence of the heterozygous state of thalassemia (a severe form of anemia) and specifically advised against marriage with another heterozygote. Premarital counseling is a potentially important aspect of genetic practice. One can imagine a stage of development in social organization and in technical methods such that prospective marital partners would be tested for heterozygosity to a considerable number of recessive genes. Even if premarital advice is not taken, it may serve a useful function of permitting early detection of disease in the offspring.

Amniocentesis, a procedure by which fluid is withdrawn from the amniotic sac surrounding the fetus, may permit prenatal diagnosis of chromosomal aberrations and of inborn errors of metabolism. Cells in the fluid can be grown in short-term culture and studied for karyotype, enzyme activities and other characteristics. In those states and countries where the laws permit, therapeutic abortion can be considered in those cases in which abnormality is detected. The focus in this approach is primarily on guaranteeing that the individual child is "well-born" and on helping parents get healthy children. Prevention of the propagation of deleterious genes and the birth of defectives (for example, those with Down's syndrome) who are prolonged financial burdens to society are secondary, although laudable, objectives. If amniocentesis with determination of the sex of the fetus were performed in all pregnancies fathered by hemophilic males and every female fetus aborted, then the frequency of the hemophilia gene would be reduced.

Genetic Factors in Common Diseases

As indicated at the outset, all disease has some genetic component in its causation. In the spectrum schematized in Fig. 9.1, common disorders, such as hypertension, atherosclerosis, and peptic ulcer, are situated toward the exogenous end. In almost all of the common disorders the genetic component in etiology is polygenic.

The genetic basis is usually quite evident in the rarer, more strictly hereditary disorders. These are now being studied in hopes of discovering the precise mode of inheritance, gene frequency, mutation rate, and dynamics in populations. The nature of the "basic" biochemical defect, enzymatic or otherwise, is also being studied. Common multifactorial disorders are being studied to find out how significant genetic factors are in the etiology and pathogenesis of the given disorder. The following approaches are used:

(1) Familial aggregation. If a disorder is genetically determined to a significant extent, then there should be more cases among the rela-

tives of persons with a given disorder than among the relatives of appropriately selected control subjects.

(2) Twin studies. Such studies provide more specific evidence on genetic etiology. In a group of twins with a given disorder, the cotwin in each case is studied to determine whether he is similarly affected or not. On the basis of the findings in the cotwin, the pair is said to be concordant or discordant. If the disorder studied is indeed genetic to a significant extent, then the concordance rate among monozygotic twins should be considerably higher than among dizygotic twins of like sex. The diagnosis of zygosity should be done by objective tests such as blood groups and other markers (p. 89).

(3) Ethnic comparisons. The comparison of frequencies of the given disorder in different ethnic groups may be a clue to the existence of genetic factors. It is not a critical method since one can never be assured of environmental comparability of the groups studied. Often the pertinent environmental factors are not even known.

(4) Blood-group and disease association. When demonstrated, this supports the existence of genetic factors; failure to find an association, however, is not evidence against such factors. The classic example is the increased risk of peptic ulcer of the duodenum in persons of blood group O and in persons who are nonsecretors. Persons who are both O and nonsecretors have a risk of peptic ulcer more than double that in the general population.

(5) The study of a possible genetic basis of component mechanisms in a given disorder may help elucidate the genetic basis of the disorder as a whole. For example, fat metabolism and disorders thereof are thought to have some relation to atherosclerosis. Genetic studies of fat metabolism are pertinent, therefore, to the genetics of atherosclerosis.

(6) Animal analogues. Analogues of human diseases occur in some animals; for example, there is a strain of rabbits with high blood pressure. Experimental breeding and other studies not feasible with humans are possible with animals.

As was indicated on p. 155, multifactorial inheritance need not mean involvement of a large number of loci. Two or three loci, with added effects of environment, can result in the characteristic features of multiple factor inheritance in families and in populations. This means that even though multifactorial inheritance is indicated by appropriate studies, if the total genetic component is strong it is worthwhile seeking the biochemical change produced by the several genes involved.

References

Carter, C. O., "The Genetics of Common Disorders," *Brit. Med. Bull.,* 25 (1969), 52–7.

Erway, L., L. S. Hurley, and A. Fraser, "Neurological Defect: Manganese in Phenocopy and Prevention of a Genetic Abnormality of Inner Ear," *Science, 152* (1966), 1766–68.

Harris, H., "Molecular Basis of Hereditary Disease," *Brit. Med. J., 1* (1968), 135–41, 1968.

Jacobson, C. B., and R. H. Barter, "Intrauterine Diagnosis and Management of Genetic Defects," *Am. J. Obstet. Gynecol., 99* (1967), 796–807. This and the paper by H. L. Nadler (below) indicate approaches to prenatal diagnosis of chromosomal and biochemical errors by study of amniotic fluid.

Jones, F. A., ed., *Clinical Aspects of Genetics.* Philadelphia: J. B. Lippincott Co., 1961.

McKusick, Victor A., "Genetics in Medicine and Medicine in Genetics," *Am. J. Med., 34* (1963), 594–99.

Nadler, H. L., "Patterns of Enzyme Development Utilizing Cultivated Human Fetal Cells Derived from Amniotic Fluid," *Biochem. Genet., 2* (1968), 119–26.

Roberts, J. A. F., *An Introduction to Medical Genetics,* 4th ed. New York: Oxford University Press, 1967.

Tatum, E. L. "Molecular Biology, Nucleic Acids, and the Future of Medicine," *Perspect. Biol. Med., 10* (1966), 19–32.

Ten

Genes and Society

The genetic composition of a population has important influences on society, and conversely social structure has important effects on the genetic make-up of a population. This interaction is nicely illustrated by the example of adult intestinal lactase deficiency and milk consumption the world over. Occasional Caucasians and a majority of Africans and Chinese have a deficiency of the intestinal enzyme lactase, which normally splits lactose (milk sugar). Milk "disagrees" with such persons; by an osmotic mechanism the undigested lactose causes abdominal discomfort and diarrhea. Early in life they have an infantile lactase and for that reason tolerate breast milk, but later when other persons develop an adult intestinal lactase, presumably by a switch mechanism similar to that from fetal to adult hemoglobin, lactase-deficient persons fail to develop lactose-digesting ability. The lack of dairying as a source of food in large parts of Africa and Asia may be owing to a genetic intolerance for milk on the part of persons living in those areas. Fermented milk may have been developed in areas of Asia and eastern Europe where the frequency of adult intestinal lactase deficiency was high. *Lactobacillus* was used to do what the intestine of the adult person could not do; thus, the genetic "defect" determined social economy.

It is possible that the interaction operated in the opposite direction—that lack of adult milk consumption led to increased frequency of the lactase deficiency gene. In areas where cattle could not be raised because of insect-borne disease or other factors, the lactase-lack would not be disadvantageous and may, for reasons unknown, have been advantageous.

Programs designed to improve the nutrition of peoples in developing countries must take account of the genetic characteristics of the populations. Powdered milk may be a biologically unacceptable food in many of these areas.

Genetic Implications of Social Forces

The impact of social forces on a population's genetic structure is evident in many other examples. Religious and social isolation can result in small endogamous groups even in large cities. The Jews have maintained a certain degree of isolation for many centuries and genetically are demonstrably distinct from the peoples among whom they have lived. Some recessive genes such as those causing pentosuria and Tay–Sachs disease occur almost only in Jews, whereas others such as PKU are rare in Jews.

Social attitudes toward miscegnation differ widely in different parts of the world. Negro-white assimilation has proceeded relatively rapidly in Brazil, where a tolerant view has prevailed. On the other hand, Negro-white mixing has proceeded more slowly in the southern United States, although after about ten generations in America the Negro race has reached the point that about 30 percent of its genes are of European derivation. Possibly the rate of Negro-white admixture in this country has been lower in the last 50 years than it was under conditions of slaveholding in the period before the Emancipation, but exact information on this point is almost impossible to obtain.

Social attitudes toward consanguinity also have important genetic implications. Although the canons of the Roman Catholic church do not entirely eliminate consanguinity among its members, there is far less consanguinity in Catholic populations than there is in populations that accept or even favor marriages among blood relations. In some endogamous Moslem groups in India as many as 40 percent of the marriages are consanguineous. All human societies—with the exception of the Egyptian pharaohs—prohibit parent-offspring and brother-sister matings; this, of course, makes good genetic sense. Other prohibitions that have prevailed in some societies, for example, that against a man's marrying the wife of a deceased brother, are genetically nonsensical—unless, of course, children affected by an autosomal recessive disorder had occurred from the first mating.

Different socioreligious attitudes toward ethnic miscegenation also existed in the past. Roman Catholic settlers of the New World— French, Spanish, Portuguese—intermarried with the American Indian and Negro, for example, in Louisiana, Mexico, and Brazil. Protestant settlers mixed much less overtly with Indian and African peoples. Most of the Protestant sects represented national religions; religious and nationalistic differences reinforced each other. The catholicity of the Roman Catholic religion appears, on the other hand, to have encouraged miscegenation.

Various social considerations are usually more important in determining the net fertility of a population than is the biological capacity of the population for reproduction. Family planning, birth control, and the prevailing fashions in family size are important factors.

In this century an important social force with potential genetic effects has been family planning. Technological developments for birth control and, in some countries (for example, Japan), legalized abortion have contributed to the force, but it was already evident previously. As a result families are smaller and more uniform in size, and fewer children are born of young and old mothers. For example, in Japan the birth rate was 35.3 per 1000 population in 1925 and 14.6 per 1000 population in 1960. The mean maternal age was 28.4 years in 1925 and 27.1 years in 1960. The standard deviation for mothers' ages was 6.7 in 1925 and 4.4 in 1960. In 1947, 64 percent of children born were of birth order 1, 2, or 3; in 1960 over 90 percent were of this lower birth order. In 1947, 2.3 percent of babies were born to mothers under 19 years of age; in 1960, the corresponding figure was 1.2 percent. In 1947, almost 20 percent of babies were born to mothers over 35 years of age; in 1960, only 5.8 percent were born to these older mothers. (These data are summarized in Table 10.1.)

The genetic effects of family planning include (1) a decline in the frequency of chromosomal aberrations such as Down's syndrome, Klinefelter's syndrome, and trisomy 18, which are related to maternal age (p. 21); (2) a decline in the rate of point mutation, which is related to paternal age (p. 40); and (3) a decline in Rh disease (p. 115), which is related to birth order. In a country such as Japan, where the level of consanguinity has been high, reduction in the size

TABLE 10.1. *Demographic Data from Japan**

Year	Standardized birth rate (per 1000 population)	Maternal age (years)		Percent live births of birth order 1, 2, 3	Percent live births to mothers	
		Mean	Variance		Under 19 years	Over 35 years
1925	35.3	28.4	45.48	—	—	—
1947	30.7	29.1	36.36	64.1	2.3	19.8
1960	14.6	27.1	19.93	90.3	1.2	5.8

* From E. Matsunaga, *J. Am. Med. Assoc., 198* (1966), 533–540.

of sibships tends to reduce the number of consanguineous matings by reducing the number of cousins available for marriage. (Because of the low frequency of consanguinity in the United States, little influence of this type is to be expected, although smaller family size must have contributed to the decline in consanguinity in the past.) A decline in the frequency of rare recessive disorders will follow on the decline in consanguinity. Of course, the frequency of the genes for these disorders will rise to a new equilibrium level when the genes are no longer lost in homozygotes. Reduction in consanguinity has in this way a dysgenic effect. It is possible that a dysgenic effect also results from the more rapid adoption of birth control among the more intelligent segments of society, but this has not been proved.

Wars and political upheavals have genetic effects of various types. An interesting rise in the sex ratio—an exaggeration of the usual excess of males over females—is observed during wartime. The sex ratio is higher in offspring born soon after marriage than it is in first offspring born late after marriage. Conception after brief contact may be accompanied by a high sex ratio; this may account for the high sex ratio in wartime.

The expulsion of the Jews from Spain in the fifteenth century and of the Huguenots from France in the fifteenth and sixteenth centuries probably had a significant effect on the genetic constitution and subsequently on the social fabric of these countries. Hitler's extermination and expulsion of Jews must inevitably have genetic and thus social effects.

Increased social and occupational mobility which is promoted by a democratic society (a "meritocracy"), combined with assortative mating, should have a beneficial effect in the long run. A free society permits the individual to find the niche in which his genetic endowment will most fully be used. Since mathematicians are more likely to associate with other mathematicians or persons of comparable competence and profession than with stevedores, assortative mating according to abilities will prevail. Such assortative mating should permit concentration of genes for intelligence and ability of many types.

In summary, social evolution and biologic evolution go hand in hand and are two aspects of an overall process of development in the human species. Social evolution awaits its Charles Darwin. It is a fascinating view that "ideas are to social evolution what genes are to biological evolution." Mutation, psychosocial selection operating on competing ideas, drift, and flow are forces in social evolution, ideas being the pertinent variable.

Social Implications of Human Genetics

Sewall Wright has suggested that one can usefully think about the social impact of a mutation in terms of a balance between con-

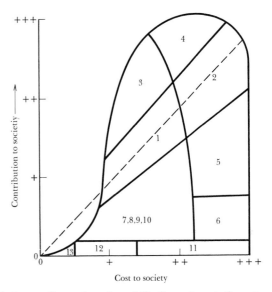

FIG. 10.1. *The relationship between the cost and contribution of mutations to society. After Sewall Wright, in* **The Biological Effects of Atomic Radiation.** **Washington, D.C.: National Research Council, 1960.**

tribution and cost. His diagram (Fig. 10.1) shows various classes of individuals according to this balance. The diagonal dotted line indicates the average ratio of contribution to cost for the population in question. The ratio may be considered 1 in a static society but as greater than 1 in a society in which there is an advance in well-being in each generation. The several groups indicated by numbers are:

(1) For the bulk of the populations, contribution and cost are balanced at a relatively modest level.

(2) In a second group, cost and contribution are also balanced but at a higher level. Professional people of average competence but with an education and standard of living well above average fall into this group.

(3) Persons who make extraordinary contributions at average cost fall into this group.

(4) In this category are persons who cost society much in terms of education and standard of living but also contribute much more than the average for their level of cost.

(5) Individuals in this class have capacities comparable to those of classes 1 through 4 but contribute less than their cost for reasons such as unearned wealth.

(6) Here Wright puts individuals of normal physical and

intellectual capacities whose cost to society outweighs their contributions because of the antisocial nature of their activities: criminals, political demagogues, etc.

In the remaining categories, cost to society definitely outweighs contribution because of physical or mental defects.

(7) Subnormal physical constitution.
(8) Low mentality but not complete helplessness.
(9) Normal to maturity but early physical breakdown by accident, infectious disease, or degenerative disorder.
(10) Mental breakdown after maturity, especially by one of the major psychoses.
(11) Complete physical or mental incapacity throughout a lifetime of more or less normal length.
(12) Death before maturity.
(13) Death *in utero* or soon after birth.

A gene responsible for placing an individual in group 13 is generally viewed as the gravest type of mutation, but its cost to society is less than that of a gene that places its bearer in some other groups, such as 10 or 11.

Nongenetic factors dominate in class 5; the role of genetic factors in class 6 is a matter of uncertainty and disagreement.

A finding in human genetics which is potentially important to social philosophy is the high order of man's normal diversity (p. 173). Although the karyotype of normal humans varies only within rather narrow limits and although fertile matings are possible among all races of man, diversity at many loci appears to be the situation in all persons. The differences between individuals in the same race are greater than the average differences between races. What matters is the individual. He should be considered on his own merits.

Social applications of genetics are illustrated by the use of genetic principles in connection with problems of disputed parentage, particularly disputed paternity. Blood groupings and other marker traits can exclude paternity but cannot establish paternity with absolute certainty, although when some unusual blood groups are found in both father and child the likelihood of paternity can reach a high level of probability. In courts of law of some states such evidence is admitted, whereas in others it is quite illogically not accepted. This legal application is, however, a trivial example of the use of genetic principles, compared to the important areas of eugenics and euphenics.

The eugenics movement had its origin with Galton before the rediscovery of Mendelism. Unfortunately the enthusiasm of the eugenicists has often run ahead of the elucidation of principles on which a sound program must be based. Eugenics can be arbitrarily classified as negative or positive. Negative eugenics is concerned with preventing reproduction of persons carrying undesirable genes. Sterilization and institutionalization are negative eugenic measures. The net result of such measures directed against those genes that are generally agreed to be undesirable is relatively minor, especially when the measures are exercised only on persons homozygous for a particular autosomal recessive gene. In many cases the problem is determining what can indeed be considered undesirable. Negative eugenics is inefficient and often the means by which it is practiced are not ethically acceptable to most people.

Positive eugenics, encouragement of reproduction by persons of presumedly superior genotype, has as many problems as negative eugenics. Hermann Muller's "germinal choice" idea is a form of positive eugenics. Muller proposed the creation of sperm banks in which the sperm of unusually capable persons would be preserved until historical perspective could validate the worthiness of the individual. The semen would then be used for artificial insemination of prospective mothers. The difficulties in identifying superior genotypes, as well as detecting deleterious recessives which may be lurking in individuals who have some superor qualities make "germinal choice" impractical. Furthermore, the undesirability of doing anything which would encourage genetic uniformity rules out the approach.

Lederberg has proposed the term *euphenics* as a counterpart of eugenics; in a sense eugenics and euphenics are comparable to the parallel terms *phenotype* and *genotype*. Eugenics concerns itself with "improvement" in the genetic material. Euphenics would concern itself with influencing the chain of information from DNA to RNA to protein to attain a more desirable phenotype. It is almost certain that in the not too distant future there will be a rapid development of methods for controlling the phenotype by influencing the gene-controlled developmental processes that are just beginning to be understood. Two areas of possible development are control of the size and capacity of the brain and modification of the immunologic barriers to organ transplantation ("spare parts" surgery) .

References

Bayless, T. M., and N. L. Christopher, "Disaccharidase Deficiency," *Am. J. Clin. Nutrition, 22* (1969) , 181–90.

Coon, Carleton S., *The Story of Man,* 2nd ed. New York: Alfred A. Knopf, Inc., 1962.

Dobzhansky, T., "Changing Man. Modern Evolutionary Biology Justifies an Optimistic View of Man's Biological Future," *Science, 155* (1967) , 409–15.

Etzioni, A., "Sex Control, Science and Society," *Science, 161* (1968) , 1107–12. The potential societal effects of the availability and application of a method for controlling one aspect of man's genetic make-up is analyzed in a thought-provoking manner.

Haller, M. H., *Eugenics: Hereditarian Attitudes in American Thought*. New Brunswick, N.J.: Rutgers University Press, 1963.

Lederberg, J., "Experimental Genetics and Human Evolution," *Amer. Naturalist, 100* (1966) , 519–31.

Matsunaga, E., "Possible Genetic Consequences of Family Planning," *J. Am. Med. Assoc., 198* (1966) , 533–40.

Mourant, A. E., and others, *The ABO Blood Groups*. Oxford, England: Blackwell Scientific Publications, 1958.

Muller, H. J., "What Genetic Course Will Man Steer?" in *Proceedings of the Third International Congress of Human Genetics (Sept. 1966)* , J. F. Crow and J. V. Neel, eds. Baltimore: Johns Hopkins Press, 1967.

National Academy of Sciences, *The Biological Effects of Atomic Radiation*. Washington, D.C.: National Research Council, 1960.

Glossary

Abiotrophy. An inborn defect which does not become clinically evident until some time after maturity. Also more generally used for disorders with late onset.

Acrocentric. Describes chromosomes with the centromere near one end so that one arm is very short.

Aggregation, familial. A concentration of cases of a disease in families. The occurrence of more cases of a given disorder in close relatives of a person with the disorder than in the families of controls.

Alleles. Alternative forms of genes which occur at the same locus. A given population may contain two or more alleles, any two of which may be carried by a given individual. The ABO system illustrates multiple allelism. Isoalleles are "normal" alleles which are distinguishable one from another only by the differences in the expression of a dominant trait produced by a gene with which they are paired in a heterozygous person.

Allograft. Graft between genetically unlike persons. Antonym: *isograft* (which see) .

Aneuploidy. Deviations from the normal diploid number of chromosomes that are not multiples of this number, for example, 45 or 47. The deviation may be of either a hypoploid or hyperploid type. See also *polyploidy.*

Anticipation. More severe expression of a disorder in a given generation than in preceding one (s) . Based largely on bias of ascertainment, anticipation has no biological basis.

Ascertainment. The manner in which individuals with a trait or disease and their families come to the investigator's attention.

Association. The occurrence together in a population of two characteristics in a frequency greater than would be predicted on the basis of chance, that is, in a frequency which is greater than the product of the frequency of each.

Assortative mating. Nonrandom mating. The preferential selection of a spouse with a particular genotype.

Assortment. The genetic consequence of the random distribution of non-homologous chromosomes to daughter cells in meiosis. See *segregation.*

Autoradiography. A technique applied in cytogenetics for the study particularly of DNA synthesis by the chromosome, using for example radioactive tritium-labelled thymidine as the marker.

Autosome. Nonsex chromosome.

Autozygous. Homozygous by descent from a common ancestor.

Backcross. Term from experimental genetics to indicate mating between heterozygote and homozygote (or between F_1 hybrid and one of the two parental strains). Double backcross is the mating between a double heterozygote and a homozygote. The mating most informative for linkage analysis is the double backcross mating involving the recessive homozygote.

Balanced polymorphism. See *polymorphism.*

Balanced translocation. See *translocation.*

Barr body, or *sex chromatin.* The chromatin mass subjacent to the nuclear membrane in cells of the normal female.

Base pair. The guanine-cytosine and adenine-thymine pairs of purine-pyrimidine bases which make up DNA. In RNA, uracil substitutes for thymine. One of the pair is on one chain, the other on the complementary chain.

Centromere, kinetochore, or *primary constriction.* The constricted portion of the chromosome, separating it into a short arm and a long arm, to which the spindle fibers are attached in mitosis and meiosis.

Chiasma formation. The cytologic basis of genetic recombination, or crossing over, occurring between homologous chromosomes at meiosis.

Chimerism. The presence in a given individual of two distinct cell types derived from different persons. The main example in man is the presence of male blood cells in the female cotwin, or vice versa. The distinction from *mosaicism* (which see) is clearest if one keeps in mind that one zygote is present at the beginning in mosaicism and two zygotes in chimerism.

Chromatid. One of the two parallel components of the metaphase chromosome joined at the centromere, each destined to form one chromosome in the daughter cell. One half of a replicated chromosome.

Chromomeres. Areas of different optical density and/or different diameters along the length of a chromosome, especially clearly discernible in prophase of meiosis.

Cistron. The smallest functional unit of genetic material, that is, that which determines the amino acid sequence of one polypeptide chain. The gene as usually conceived is synonymous with the cistron.

Clone.   Cells all derived from a single cell by repeated mitosis and all having the same genetic constitution.

Codominance.   When both alleles are expressed in the heterozygote.

Codon.   The triplet of nucleotides (DNA) specifying a particular amino acid.

Colinearity.   The relationship between two macromolecules (DNA and protein), in which the sequence of components (bases) of the former specifies the sequence of components (amino acids) of the latter.

Compound.   Person bearing two different mutant alleles at a given locus, for example, a person with both hemoglobin S and hemoglobin C. Not to be confused with the doubly heterozygous state.

Concordance.   Usually used in reference to twins, to indicate that both have a given trait. *Discordance* is the antithetical term.

Congenital.   Present at birth. No necessary connotation as to genetic or nongenetic causation.

Consanguinity (inbreeding), coefficient of.   Probability of autozygosity (homozygosity by descent from a common ancestor).

Consultand.   Person seeking genetic counsel, usually the person whose genotype is in question.

Coupling.   In a double heterozygote, linkage is said to be in the coupling phase when the mutant alleles of interest at the two loci are on the same chromosome. See *repulsion,* the antithetical term.

Crossing over.   See *recombination.*

Cytogenetics.   The branch of genetics concerned mainly with the chromosomes and correlations with the phenotype.

Deletion.   Loss of part of a chromosome.

Deme.   A consanguineal kin group, or local endogamous community.

Dictyotene.   The interphase-like stage in which the oocyte persists from late fetal life until ovulation. The oocyte has not yet completed the first stage of meiosis.

Diploid.   The double state of all chromosomes in normal somatic cells (Symbol: 2*N*.) Compare *haploid.*

Dizygotic (dizygous).   Refers to twins derived from separate eggs ("fraternal twins").

Dominant.   The attribute of a trait which is expressed when the responsible gene is present in heterozygous state, or "single dose." The phenotype in the homozygote may or may not differ from that in the heterozygote.

Dosage compensation.   The mechanism by which the two X chromosomes of the normal female are rendered quantitatively identical in their effect to the one X chromosome of the normal male. See *Lyon hypothesis.*

Drift.   See *random genetic drift.*

Endogamy.   Mating within the group. Synonym: *inbreeding.*

Endoreduplication.   A process in which the chromosomes replicate without cell division.

Epistasis.   A form of interaction between genes (products) situated at different loci, where one gene masks or prevents the expression of another gene at a different locus. (Dominance is the term used when one gene masks one of its own alleles, that is, another gene at the same locus.)

Euchromatin.   See *heterochromatin.*

Expressivity. The variability in severity of a genetic trait.

Family. See *kindred.*

Fitness, Darwinian. Measured by reproductive performance, that is, number of offspring. The average is considered a fitness of 1.

Forme fruste. An incomplete, partial or mild form of trait or syndrome.

Gene flow. Transfer of genes from one population to another by migration and miscegenation.

Genetic death. Failure of the individual to reproduce.

Genetic load. See *load, genetic.*

Genocopy, or *genetic mimic.* See *heterogeneity.*

Genome. The total genetic endowment.

Genotype. The genetic constitution, either at one specific locus or more generally. In the general sense, genotype is essentially synonymous with genome. See *phenotype.*

Gonadal mosaic. See *mosaicism.*

Gonosome. Term now rarely used, referring to sex chromosomes. Compare *autosome.*

Haploid. The single state of all chromosomes, normal in gametes. (Symbol: 1N.)

Hardy–Weinberg law. If two alleles (*A* and *a*) occur in a randomly mating population with the frequency of p and q, respectively, where $p + q = 1$, then the expected proportions of the three genotypes $AA = p^2$, $Aa = 2pq$, and $aa = q^2$, remain constant from one generation to the next. Mutation, selection, migration, and genetic drift can disturb the Hardy–Weinberg equilibrium.

Hemizygous (hemizygotic). The state of the male with regard to the X chromosome.

Hereditary, heritable, and *heredofamilial.* Essentially synonymous terms for transmissible genetic traits.

Heterochromatin. Chromosomal material which differs in its staining properties from the rest. Example: the X chromosome which constitutes the Barr body in the normal female. Antonym: *euchromatin.*

Heterogametic. Capable of producing two kinds of gametes. In man the male is the heterogametic sex since X-bearing and Y-bearing gametes are formed.

Heterogeneity. An identical or similar phenotype with different genetic bases. Genocopy and genetic mimic are terms for a genetic trait which is phenotypically similar to but fundamentally distinct from another.

Heterozygous (heterozygote). Possessing different alleles at a given locus. *Doubly heterozygous* (or *double heterozygote*) refers to heterozygous state at two separate loci. The SC individual is not properly referred to as doubly heterozygous since the S and C genes are allelic; *double mutant* or *compound* may be a more accurate designation.

Holandric. Literally, affecting all males. Applied to Y-linked inheritance. "Hairy pinnae" is one of the few traits in which holandric inheritance has strong support, but even this is not completely conclusive.

Homologous chromosomes. The chromosomes of a pair in the diploid complement.

Homozygous (homozygotic). Possessing identical genes at a given locus.

Hyperploidy, hypoploidy.   See *aneuploidy.*

Ideogram.   A sketch representing the karyotype.

Immunogenetics.   The genetics of immunity, including the immune globulins, blood groups, and histocompatibility.

Interchange.   See *translocation.*

Interphase.   Improperly called "resting phase." The interval between cell division.

Inversion.   End-to-end reversal of a segment within a chromosome. When in one arm of a chromosome it is referred to as paracentric; when the inverted segment includes the centromere, the inversion is termed pericentric.

Isoalleles.   See *alleles.*

Isochromosome.   An anomalous chromosome, with median centromere and identical arms, which arose through transverse splitting of the centromere rather than longitudinal splitting between two chromatids.

Isograft.   Graft between genetically identical individuals, in man monozygotic twins. Antonym: *allograft* (which see) .

Isolate.   A genetically separate endogamous population. See also *deme.*

Karyotype.   The chromosomal constitution. A term parallel to genotype and phenotype.

Kindred.   Family in the larger sense. The term family is usually restricted to the nuclear family (parents and children) .

Kinetochore.   See *centromere.*

Lethal equivalents.   See *load, genetic.*

Linkage.   Location of genes on the same chromosome.

Load, genetic.   Deleterious recessive genes carried in heterozygous state. Five genes which in the homozygote each reduces the fitness by 20 percent make one lethal equivalent. Genetic load is measured by observing the effects of inbreeding.

Locus.   The site of a chromosome occupied by a specific set of alleles.

Lyon hypothesis (Lyon phenomenon or principle) .   The genetic inactivation of all X chromosomes in excess of one, on a random basis in all cells at an early stage of embryogenesis.

Map unit, or *map distance.*   The measure of distance separating loci on a chromosome as inferred from the observed recombination fraction. For closely situated loci the map distance is essentially identical to the recombination fraction.

Meiosis.   The process occurring during gametogenesis by which the diploid chromosome number is reduced to the haploid number of the sperm and ova. The word is the same as *miosis,* meaning a reduction in size. By convention miosis is used for small pupils and meiosis for the genetic process.

Mendelizing.   The attribute of traits which show simple patterns of inheritance.

Metacentric.   Refers to chromosomes with the centromere near the middle.

Metaphase.   The stage of cell division when the contracted chromosomes, each consisting of two chromatids, are arranged on the spindle plate.

Mitosis.   The process of cell division in somatic tissue, during which chromosome replication and division insures a normal diploid number in each of two daughter cells.

Monomeric. Determined by gene (s) at a single locus, either in heterozygous or homozygous state.

Monosomy. The state which obtains when only one of a given chromosome is present rather than the normal two. The only example in man: XO Turner syndrome. Partial monosomy results when part of a chromosome is deleted so that that portion of the homologue is monosomic.

Monozygotic (monozygous). Refers to twins derived from one egg ("identical twins").

Mosaicism. Presence in the same individual of two or more distinct but related cell populations, the different cell types not arising by a process of neoplasia or chimerism (which see) but rather one or more types of cell rising from a single cell type either by gene mutation or chromosomal aberrations. Gonadal mosaicism results from a somatic mutation early in embryogenesis so that part or all of the germ cells are of mutant type. A person with a gonadal mosaicism for a dominant trait can have two or more children with that trait without showing it himself. Gonadal mosaicism may also involve a chromosomal aberration.

Multifactorial. Determined by multiple genetic and nongenetic factors. Polygenic is the term used to describe multiple genetic factors specifically.

Mutation. Change in the genetic material. Usually refers to point mutation, that is, change in a single gene, but in a more general sense includes changes recognizable microscopically as chromosomal aberrations. In connection with inherited diseases, mutation in the germ cells are most relevant, but somatic mutation also occurs and may be the basis of some cancers and some aspects of aging.

Mutation rate. The number of mutations which occur at a particular locus per gamete per generation.

Muton. The smallest mutational unit of genetic material. Since change in a single base changes the genetic code, the muton is equivalent to one base.

Nondisjunction. Failure of two homologous chromosomes to go to separate cells in the first stage of meiosis or failure of the two chromatids of a chromosome to pass to separate cells in the second stage of meiosis or in mitosis. Secondary nondisjunction is illustrated by the formation of gametes with two chromosomes 21 by persons with trisomy 21. These can give rise to offspring with Down's syndrome.

Nonpenetrance. Failure of a genetic trait to be evident even though the genotype usually productive of that phenotype is present.

Nonsecretor. See *secretor.*

Nucleolus. Vacuole-like body, rich in RNA, in the nucleus.

Nucleolus organizer. Secondary constrictions of chromosomes, particularly those related to satellites, seem to have this function.

Nucleoside. The combination of a purine or pyrimidine base and a sugar.

Nucleotide. The combination of a purine or pyrimidine base, a sugar, and a phosphate group.

Oogenesis. The process of formation of the female gametes.

Operon. A group of closely linked structural genes with related function which are turned on and off in concert under the control of a postulated controlling gene, or operator.

Panmixis. Random mating.

Penetrance. See *nonpenetrance.*

Pharmacogenetics. That part of genetics concerned with the relationship between genotype and drug effects.

Phenocopy. An environmentally induced mimic of a genetic disorder.

Phenotype. The observable characteristics of the organism. The distinction between genotype (which see) and phenotype is similar to that between character and reputation.

Pleiotropism. Loosely, "many effects." Many hereditary syndromes are examples of the pleiotropic effect of a single gene. Polyphenic is a synonym for pleiotropic, parallel with polygenic.

Point mutation. See *mutation.*

Polygenic. See *multifactorial.*

Polymorphism or *genetic polymorphism.* "The occurrence together in the same habitat of two or more discontinuous forms of a species in such proportions that the rarest of them cannot be maintained merely by recurrent mutation" (Ford, 1940). When the polymorphism is maintained through a selective advantage of the heterozygote, it is referred to as balanced polymorphism.

Polyphenic. See *pleiotropism.*

Polyploidy. Presence of chromosomes in numbers which are simple multiples of the normal diploid number. Tetraploid = 92 chromosomes; octaploid = 184 chromosomes, and so on. Triploid = 69 chromosomes, a frequent finding in abortuses. See *aneuploidy.*

Primary constriction. See *centromere.*

Proband. The affected individual who first comes to attention and brings the family to study.

Propositus (female, *proposita;* plurals, *propositi* and *propositae*) : Synonyms are *proband* (which see) or *index case.* (*Proband* is preferred.)

Quasi-dominance. The direct transmission, generation to generation, of a recessive trait if the gene is frequent or inbreeding is intense.

Random genetic drift, or simply *drift.* Chance variation in gene frequency from one generation to another. The smaller the population, the greater are the random variations.

Recessive. The attribute of a trait which is expressed *only* when the responsible gene is present in homozygous state, or "double dose." Subtle differences may be detectable in the heterozygote, however.

Recombination and *crossing over.* The physical process by which genes at loci on the same chromosome end up on separate chromosomes in the subsequent generation (s) is crossing over. Recombination is the consequence of crossing over: the change from coupling to repulsion, or vice versa.

Recon. The smallest unit of recombination. Crossing over can occur within the cistron. The smallest unit of recombination is between successive nucleotides in a codon.

Repulsion. In a double heterozygote, linkage is said to be in the repulsion phase when the mutant alleles of interest at the two loci are on opposite chromosomes.

RNA. Ribonucleic acid, one type of which, messenger RNA, is the intermediary between DNA and the protein synthesizing machinery in the cytoplasm. Other types are ribosomal RNA and transfer (or soluble) RNA.

Satellite. A globoid chromatin mass attached through a secondary constriction to the end of a chromosome.

Secondary constriction. Narrowed heterochromatic area in a chromosome. The centromere is the primary constriction. A secondary constriction separates the satellite from the rest of the chromosome. Probably associated with nucleolus formation. See nucleolus organizer.

Secondary nondisjunction. See nondisjunction.

Secretor. The secretion of ABO blood group substance into saliva and body fluids.

Sectorial. Presence of, or individual with, sector of tissue which carries a somatic mutation (which see), and which is therefore different phenotypically from the rest of the tissues. If the sector is gonadal, this is referred to as *gonadal mosaicism* (which see).

Segregation. The genetic consequence of separation of homologous chromosomes in meiosis. See *assortment.*

Selection. Differential net reproductive performance according to genotype.

Sex chromatin. See *Barr body.*

Sex limitation, or *sex influence.* These terms are applied to autosomal traits which for some reason occur only in either males or females (sex limitation) or predominantly in one sex (sex influence).

Sex-linked. Determined by a gene located on the X chromosome. Since a Y-linked trait is also sex-linked, X-linked is a preferable term.

Sex ratio. The ratio of males to females. The primary sex ratio refers to that at fertilization; the secondary sex ratio to that at birth.

Sibs, siblings. Brothers and sisters. Brevity makes *sib* the preferred term.

Sibship. Group of brothers and/or sisters.

Somatic mutation. See *mutation.*

Spermatogenesis. The process of formation of spermatozoa.

Spermiogenesis. That part of spermatogenesis in which spermatids develop into spermatozoa.

Structural gene. One which specifies the amino acid sequence of a polypeptide chain.

Synapsis. The process by which homologous chromosomes come to lie side-by-side early in meiosis.

Syndrome. The association of phenotypic manifestations.

Telocentric. Refers to chromosome with its centromere at the end.

Tetraploid. See *polyploidy.*

Trait. Any gene-determined characteristic. Although in medicine it has come to be used particularly for the heterozygous state of a recessive disorder such as sickle cell anemia, it has a more general meaning in genetics.

Transcription. Transfer of the genetic code information from DNA to messenger RNA. See *translation.*

Transduction. Change in the genetic constitution of a cell by transfer of DNA from a virus to the genome of that cell.

Translation. Transfer of the genetic information carried by messenger RNA into the amino acid sequence of proteins. See *transcription.*

Translocation. The displacement of part or all of one chromosome to another. When an individual or gamete carries no more or less than the normal diploid or haploid genetic material, respectively, the situation is referred to as *balanced translocation. Interchange* is a suggested alternative term for translocation.

Triploidy. See *polyploidy.*

Trisomy. The state which obtains when three of a given chromosome are present rather than the normal two. Example: trisomy 21 (Down's syndrome).

Wild type. The normal allele of rare mutant gene, sometimes symbolized by +.

Zygote. The fertilized ovum.

A Guide to Further Reading

General Sources

Boyer, S. H., IV, ed., *Papers on Human Genetics*. Englewood Cliffs, N.J.: Prentice-Hall, Inc., 1963.

Carter, C. O., *Human Heredity*. Baltimore: Penguin Books, Inc., 1962.

Crow, J. F., and J. V. Neel, eds., *Proceedings of the Third International Congress of Human Genetics (Sept. 1966)*. Baltimore: Johns Hopkins Press, 1967.

Ford, C. E., and H. Harris, eds., "New Aspects of Human Genetics," *Brit. Med. Bull.*, 25 (Jan., 1969), No. 1, 118 pp.

McKusick, V. A., *Mendelian Inheritance in Man. Catalogs of Autosomal Dominant, Autosomal Recessive and X-linked Phenotypes*, 2nd ed. Baltimore: Johns Hopkins Press, 1968.

Moody, P. A., *Genetics of Man*. New York: W. W. Norton & Co., Inc., 1967.

Penrose, Lionel S., *Outline of Human Genetics*, 2nd ed. London: William Heinemann, Ltd., 1963.

Roberts, J. A. F., *An Introduction to Medical Genetics*, 4th ed. New York: Oxford University Press, 1967.

Stern, Curt, *Principles of Human Genetics*, 2nd ed. San Francisco: W. H. Freeman & Company, Publishers, 1960.

Sutton, H. E., *An Introduction to Human Genetics*. New York: Holt, Rinehart & Winston, 1965.

Thompson, J. S., and M. W. Thompson, *Genetics in Medicine*. Philadelphia: W. B. Saunders Co., 1966.

Biochemical Genetics

Garrod, Archibald E., *Inborn Errors of Metabolism.* Reprinted with a supple-
ment by Harry Harris. New York: Oxford University Press, 1963.
Hsia, D. Y.-Y., *Inborn Errors of Metabolism,* 2nd ed. Chicago: Year Book
Medical Publishers, 1966. (See Part 2 for methods of screening for
inborn errors of metabolism and other genetic biochemical defects.)
Stanbury, J. B., J. B. Wyngaarden, and D. S. Fredrickson, eds., *The Metabolic
Basis of Inherited Disease,* 2nd ed. New York: McGraw-Hill Book
Co., 1966.

Human Cytogenetics

Montagu, M. F. A., ed., *Genetic Mechanisms in Human Disease: Chromosomal
Aberrations.* Springfield, Ill.: Charles C Thomas, Publisher, 1961.
Turpin, R., and J. Lejeune, *Les chromosomes humains (caryotype normal et
variations pathologiques)* . Paris: Gauthier-Villars, 1965.
Yunis, J. J., ed., *Human Chromosome Methodology.* New York: Academic
Press, Inc., 1965.

Statistical Genetics

Li, C. C., *Human Genetics.* New York: McGraw-Hill Book Company, 1961.

Periodicals

American Journal of Human Genetics (published every two months by the
American Society of Human Genetics) .
Annals of Human Genetics (published quarterly by the Galton Laboratory,
London) .
Human Heredity (formerly, *Acta Genetica et Statistica Medica*) .
Journal of Medical Genetics (a publication of the British Medical Associ-
ation) .
Steinberg, Arthur G., and Alexander G. Bearn, eds., *Progress in Medical Ge-
netics.* New York: Grune & Stratton, Inc., 1961 (Vol. I) , 1962 (Vol.
II) , 1964 (Vol. III) , 1965 (Vol. IV) , and 1967 (Vol. V) . A collection
of long reviews in specific areas of human genetics.

Subject Index

Author Index